Melone
Chicago
'94

BONE HUNTERS IN PATAGONIA

BAND OF GUANACO.

BONE HUNTERS IN PATAGONIA

NARRATIVE OF THE EXPEDITION

J. B. HATCHER

OX BOW PRESS

WOODBRIDGE, CONNECTICUT 06525

Reprinted from the Reports of the Princeton University Expeditions
to Patagonia, 1896–1899.

Ox Bow Press
P.O. Box 4045
Woodbridge, CT 06525

Library of Congress Cataloging in Publication Data

Hatcher, J. B. (John Bell), 1861–1904.
  Bone hunters in Patagonia.

  Reprinted from: Princeton University expeditions to
Patagonia, 1896–1899. Princeton: Princeton University,
1903.
    1. Vertebrates, Fossil.   2. Paleontology—Patagonia
(Argentina and Chile)   I. Title.
QE841.H29   1985   566'.0982'7   84–25527

ISBN 0–918024–36–6 Hardcover
ISBN 0–918024–37–4 Paperback

Printed in the United States of America

TO THE MEMORY

OF

OTHNIEL CHARLES MARSH

STUDENT AND LOVER OF NATURE

THIS VOLUME IS DEDICATED

BY

THE AUTHOR

FOREWORD TO 1985 REPRINT

DURING the years 1896–1899, J. B. Hatcher, accompanied by O. A. Peterson and A. E. Colburn, conducted one of the world's most successful fossil hunts through the southern tip of South America known as Patagonia. The enormous collections he amassed were sent to Princeton University where they were studied by scholars for many years and still form part of the collection. The expeditions were difficult and Hatcher showed indomitable determination and courage in the face of many obstacles.

The scientific results of Hatcher's works appear in a massive seven volume series issued in 1903. Within the first of these volumes is the *Narrative of the Expeditions,* a scientific adventure story that can best be compared with the South American sections of Darwin's *Voyage of the Beagle.* Yet Hatcher's story has been unknown except to scholars who penetrated the forbidding *Reports of the Princeton University Expedition to Patagonia 1896–1899.* In order to bring Hatcher's fascinating tale to a broader public the narrative portions are here reproduced in a facsimile edition. This is one of the great exploration stories of all time and we are pleased to bring it back to life after eighty years in library storage.

THE EDITORS.

WOODBRIDGE, CONNECTICUT 1985.

EDITOR'S PREFACE.

THE great enterprise of a scientific exploration of Patagonia, which was planned by Mr. J. B. Hatcher, resulted in a remarkable success, because of the indomitable determination in the face of most discouraging obstacles, which was displayed by him and by Messrs. O. A. Peterson and A. E. Colburn, his two assistants. The reader of Mr. Hatcher's "Narrative" will gain some conception of the difficulties of every kind with which the work was beset, but one must read between the lines to understand how great those difficulties were and how often they seemed to be insurmountable. Only the greatest courage, in union with long experience and unusual skill, could have achieved such distinguished success.

It should be emphasized that the chief object of the expeditions was to make collections of the vertebrate and invertebrate fossils of Patagonia, in which the discoveries of the brothers Ameghino had so strongly aroused the interest of the scientific world. Some of the most important and far-reaching of geological and biological problems had been raised by the writings of Dr. Florentino Ameghino and it seemed most desirable to have a thoroughly representative series of the Patagonian fossils in some museum where they might be minutely studied in connection with the fossils of the northern hemisphere. In this, their principal purpose, the expeditions were brilliantly successful. How large and important the collections are, may be readily seen from an examination of the palæontological monographs of these Reports, Volumes IV. to VII. inclusive. It is confidently expected that the study of this abundant material (still far from completion at the present date of writing) will bring the definitive solution of some of the problems alluded to above.

The intelligent collection of fossils involves the determination of stratigraphical succession, and Mr. Hatcher was able to accomplish a great deal of most useful work in this connection, making possible, for the first time, a rational account of the geology of large areas in southern Patagonia.

He discovered a number of previously unknown geological formations and has made clear the order of succession of the various Tertiary horizons. As the pursuit of this object led the party into unexplored regions, many geographical discoveries resulted, the importance of which has not been diminished by the subsequent and much more elaborate surveys of the Argentine and Chilian Boundary Commissions.

While the collection of fossils and the determination of stratigraphy were, as already stated, the main objects in view, zoölogy and botany were by no means neglected, and very valuable collections of the recent mammals, birds and plants were secured from the wide regions explored, together with smaller, but still most useful series of the amphibians and fresh-water fishes, of the molluscs and other invertebrate groups. Much that was new to science was thus gathered and the whole forms a monument of energy and skill, which it is difficult to characterize without using terms that seem to savour of exaggeration.

This additional material, as it may be called, is described in Volumes II., III. and VIII. of the Reports.

An undertaking of such magnitude could not have been carried out without the interest and coöperation of many friends, both at home and in South America, and it is a pleasant duty to express here our grateful thanks to those whose help made success possible. The expenses of the expeditions were borne by many graduates and friends of Princeton University, who have supported several enterprises of a similar kind, though on a more modest scale. In particular, thanks are due to the following gentlemen, who rendered indispensable services, both by contributing to the fund and by giving freely of their time and labors to the support of the work: Messrs. John W. Garrett and the late Horatio W. Garrett, of Baltimore; Mr. Henry G. Bryant, of Philadelphia; Messrs. J. B. Hatcher, (now of Pittsburgh), Charles W. McAlpin, Junius S. Morgan and M. Taylor Pyne, of Princeton; Messrs. John W. Aitken, James W. Alexander, C. Ledyard Blair, Arthur A. Brownlee, John L. Cadwalader, Cornelius C. Cuyler, Cleveland H. Dodge, Harvey Edward Fisk, R. T. H. Halsey, Parker Handy, Morris K. Jesup, Rollin H. Lynde, John J. McCook, Henry Fairfield Osborn, William C. Osborn, Stephen S. Palmer, James Tolman Pyle, W. Scott Pyle, Percy R. Pyne, Philip A. Rollins, Rudolf Schirmer, Charles Scribner, Francis Speir, Jr., and M. Allen Starr, of New York; Messrs. David B.

Jones, Thomas D. Jones, Cyrus McCormick and Stanley McCormick, of Chicago.

Several officers of the United States Government took a very helpful interest in the expeditions and rendered invaluable aid, making the way plain before them and securing for them the recognition of the Argentinian officials, of whose courtesy Mr. Hatcher speaks gratefully. Hon. James Wilson, Secretary of Agriculture; Hon. David J. Hill, Assistant Secretary of State; Hon. George Gray, Justice of the United States Supreme Court; Dr. C. Hart Merriam, of the Department of Agriculture, and Mr. W J McGee, of the Bureau of Ethnology, have a large claim upon our gratitude for repeated and most valuable assistance.

Appreciative mention should also be made of the kindness of Messrs. Busk and Jevons, New York agents of the Lamport & Holt line of steamers, and of Messrs. W. R. Grace & Co., who aided the expeditions in very material fashion in the matter of passenger fares and freight charges on their steamships.

In the spring of last year (1901), the editor of these volumes found it necessary, in the interest of the publication, to visit the Museums of La Plata and Buenos Aires and study the collections there gathered. He is glad of this opportunity to express his feelings of profound gratitude to those who did everything in their power to render these investigations helpful and satisfactory.

Dr. Florentino Ameghino, now Director of the National Museum at Buenos Aires, but then living at La Plata, permitted the freest possible use of his great private collection of Patagonian fossils, a collection which is especially valuable because it contains by far the largest number of the type-specimens of the genera and species named from Patagonian formations.

Dr. F. P. Moreno, Director of the La Plata Museum, then in London as a member of the Argentinian Boundary Commission, placed his museum and all its resources unreservedly at the disposal of the visitor, who, living in the building for several months, enjoyed its hospitality in the most literal sense of the word. The Secretary of the Museum, Sr. R. Catani, and the Curators, Dr. Santiago Roth, Professor R. Hauthal, Dr. R. Lehmann-Nitsche and Sr. Carlos Bruch, were indefatigable in their assistance, and most hearty thanks are due to each and all of them for countless instances of helpful kindness.

In Buenos Aires, the late Dr. C. Bergh, then Director of the Museum, and Sr. R. Pendola, Secretary, were extremely courteous and provided every facility that the Museum could offer.   The United States Minister to Argentina, Hon. William P. Lord, also interested himself in the enterprise and assisted it in every way that he could command.

The editor wishes also to express his sense of gratitude to the many collaborators, both at home and abroad, who have taken part in the laborious task of describing the various parts of the collections which have been entrusted to their hands.   To their scientific colleagues the names of these collaborators are a sufficient guarantee of the quality of the work.

Finally, most grateful mention must be made of the generosity of J. Pierpont Morgan, Esq., who has rendered it possible to publish these reports in a manner altogether worthy of their subject and with adequate illustrations.   Scientific undertakings of this magnitude are generally executed by governments and are beyond the scope of universities. Mr. Morgan's liberality alone has put it in our power to bring together in one uniform series all the great and varied results of Mr. Hatcher's labors in South America.   In expressing his most sincere thanks for this great gift, the editor acts as spokesman for a numerous and widely spread body of students, as well as for Princeton University.

WILLIAM B. SCOTT.

PRINCETON, N. J., December 8, 1902.

Bone Hunters in Patagonia

NARRATIVE

OF THE

PRINCETON UNIVERSITY EXPEDITIONS TO PATAGONIA,

MARCH, 1896, TO SEPTEMBER, 1899.

INTRODUCTION.

SINCE the publication of the discoveries and observations of Darwin, who, as naturalist to the "Beagle," visited Patagonia in 1833–36, there has existed among naturalists everywhere an intense interest in the natural history of that region. This could hardly have been otherwise, for whatever subject or country then claimed the attention of that master mind became at once invested with a peculiar interest, through the links they thus formed in that long chain of observations which led to those broad generalizations and deductions that, when advanced a few years later, were so startling as to provoke, by their very boldness, universal attention. This attention at first took the form, in most instances, of adverse criticism, even among scientists of repute. There were a few notable exceptions like Wallace, Huxley, and Haeckel, who were not slow to champion the theories advanced by Darwin, and the latter lived to see the conclusions at which he had arrived with such painstaking care universally accepted, even by that class of people who had at first met them with ridicule and hurled at their author opprobrious epithets, for want of more convincing arguments with which to refute his well-founded doctrines.

Darwin's interest in Patagonia, as in the other countries visited, was of the broadest nature, and his account of the geology and natural history of such parts of the region as were visited by him is still and always will be a standard work, remarkable for its completeness and accuracy, considering the means at his command and the pioneer nature of the work.

Not the least interesting of the discoveries made by Darwin on this voyage were the remains of extinct fossil vertebrates which he found imbedded in the rocks of the sea cliffs at various points along the eastern coast, and more especially at Bahia Blanca, San Julian, and the port of Gallegos.[1] These were later studied and described by Professor Richard Owen, and from his descriptions of the rather meagre remains brought home by Darwin palæontologists received the first intimation of that entirely new world of animal life, which, though for the most part now quite extinct, was in comparatively recent times, from the standpoint of the geologist, extremely rich. The bones of these animals were buried in the muds and sands that accumulated about the margins and over the beds of prehistoric rivers, lakes and swamps. These sands and mud flats were later solidified and now form the different strata of sandstones and shales that constitute the cliffs of the sea and the bluffs of the streams which have eroded their beds deep into the surface of the Patagonian plains. In these rocks are still preserved numerous skulls and skeletons of the animals that inhabited this region in those prehistoric times, and their remains in varying states of preservation may now be seen lying in great abundance at various localities upon the bare and freshly eroded surfaces of the hills, or protruding from the face of the rocky escarpment that extends almost continuously along the coast, and from the slopes of the bluffs of all the more important water courses throughout the interior.

The novelty and wealth of this extinct fauna were fairly indicated by the discoveries of Darwin and at the time aroused considerable interest, but, strangely enough, this was allowed to subside, and a half century passed before any but a casual interest was taken in exploring and developing the marvelous wealth of this newly discovered treasure-house, rich in the remains of prehistoric life, for as yet no Marsh or Cope had arisen in South America to give the needed impetus to the work. It is true that Burmeister and one or two others did something toward increasing our knowledge of the geology and palæontology of this region, but their work had for the most part been done in an unsystematic and per-

[1] Darwin himself did not visit Gallegos. The discovery of vertebrate fossils at this port was made a few years earlier by Captain Sullivan, but has been referred to by Darwin.

functory manner and no determined effort was put forth toward bringing to light the abundant wealth of the region in the remains of extinct life. In 1887 Señor Carlos Ameghino accompanied an expedition to southern Patagonia, and with that year there commenced that series of discoveries which have followed one another in such rapid succession as a result of the vigorous manner in which the work has been pursued by the brothers Carlos and Florentino Ameghino.

The discoveries of the Ameghinos were of such importance as to arouse the interest of palæontologists and geologists everywhere. Interesting and remarkable as were their discoveries (for it was really a new world of animal life that was being brought to light, totally unlike anything hitherto known either among living or fossil faunas) quite as startling were many of the theories advanced by Dr. Florentino Ameghino, concerning the age of the various beds and the relation of the fauna to certain extinct and living animals of the northern hemisphere. For several years geologists and palæontologists everywhere had realized the importance of the work being carried on by the Ameghinos, though at the same time recognizing the necessity of making a thorough study of the Tertiary and Cretaceous deposits of Patagonia together with their contained fossils, in accordance with the more careful and painstaking methods which have been developed in the northern hemisphere during a half century by a great number of trained and skilled observers, belonging to two generations. It was believed that, when the light of all that had been discovered bearing upon the geological sequence and development of animal life as worked out in the northern hemisphere had been thrown with its full force upon those of the southern, many of the apparently conflicting observations and theories set forth by the Ameghinos would prove invalid, while the main facts would be found to harmonize with those already well established in the north. It was for this purpose that the Princeton University expeditions to Patagonia were organized and carried out by the present writer, when Curator of the Department of Vertebrate Palæontology of that institution.

While the primary object of the undertaking was to make observations and collections bearing upon the geology and palæontology of the region, such attention as was possible, considering our limited equipment, was given to other branches of natural history. Without wishing to apologize in any way for the character or magnitude of the work accomplished, of the value of which the interested investigator will draw his own conclusions, based upon the contents of this and the other volumes of the Reports, always remembering, it is hoped, that the personnel of the expeditions never consisted of more than two persons, I should like to emphasize the fact that the expeditions were essentially palæontological. There were three of these expeditions. The first extended from March 1, 1896, to July 16, 1897, and consisted of the writer and Mr. O. A. Peterson as assistant. The second extended from November 7, 1897, to November 9, 1898, and was composed of the writer, with Mr. A. E. Colburn as taxidermist. The third was carried on from December 9, 1898, to September 1, 1899, when Mr. O. A. Peterson again accompanied the writer as assistant.

The Expeditions were planned and carried out by the present writer. Professor W. B. Scott, as head of the Department of Geology and Palæontology in the University, gave freely his influence and best efforts toward their accomplishment. The undertaking received the moral support, much encouragement and substantial financial assistance from various friends and alumni of the University and from others interested in the promotion of our knowledge of natural history.

The following gentlemen may be mentioned as among the chief contributors to the first two Expeditions : Messrs. John W. Garrett, H. W. Garrett,[1] M. Taylor Pyne, C. H. Dodge, F. Speir, C. C. Cuyler, Morris K. Jesup, P. A. Rollins. To each of these the writer wishes to extend his best personal thanks, and at the same time improve the opportunity to accord to them in these volumes that public recognition which their generosity has so justly merited. The expenses of the third expedition were for the most part met by the present writer.

Much valuable assistance was received from the United States Bureau of Ethnology and the Department of Agriculture at Washington, from each of which honorary appointments were given the writer, which secured for the expeditions official recognition in the countries visited and thus greatly facilitated the work. In this connection we were under special obligations to Mr. W J McGee, who was ever ready to aid us in every way possible. Our thanks are also due to Dr. C. Hart Merriam, of the Department of Agriculture.

To the Lamport and Holt and the W. R. Grace S. S. lines we were indebted for reduced rates of passage to and from South America.

[1] Since deceased.

The Argentine officials in Buenos Aires were extremely kind and furnished us with free transportation on their transport boats from that city to the various ports of call on the coast of Patagonia and return.

At Gallegos, the capital of the territory of Santa Cruz and our headquarters while in Patagonia, we were made especially welcome by Governor Edelmiro Mayer and his amiable wife, and given every assistance possible by the genial secretary, Señor Juan D. Aubone, and the chief of police, Señor Cornelius Villegrand.

Last, but by no means of least importance to the final success of the undertaking was the universal and kindly hospitality shown us by the Patagonian *estancieros* wherever met with. The generous assistance of these homely people contributed much to our success, and the names of Halliday, Felton, Rudd, Montes, Fernandez, Kyle, McCoy and others will long be remembered by us, reviving memories of welcome and good cheer in the midst of the interesting, but, to say the least, somewhat cheerless three years, passed as wandering naturalists on the Patagonian plains.

To Mr. W. E. D. Scott the writer is especially indebted for the correct identification of the birds mentioned in the present volume, while Dr. George Macloskie has identified a number of the plants and Dr. A. E. Ortmann has identified for me a number of the crustacea mentioned in the Narrative. The illustrations in this volume are from photographs by the author.

CHAPTER I.

Embark at Martin's Pier, Brooklyn; Delayed by fog in the lower harbor; Rough weather while crossing the Gulf Stream; Through the Sargasso Sea; Flights of flying fish; The Portuguese Man of War; Phosphorescence of surface waters in the Tropics; Sighting the Pernambuco lights; Lobos Island; Maldonado; Montevideo, its architecture and cleanliness; Arrival at Buenos Aires; Courteous treatment from the Argentine officials; A visit to the La Plata Museum; Leave Buenos Aires for Gallegos; The River Plate; Bahia Blanca; Argentine soldiers; San Blas; New Bay; Port Desire; Santa Cruz; Arrival at Gallegos.

EARLY on the morning of February 29, 1896, we embarked at Martin's Pier, Brooklyn, on the S. S. Gallileo of the Lamport and Holt line, which was to convey us as far as Buenos Aires. Surely the name of so renowned a scientist as was Gallileo should be considered a good omen, if not an absolute guarantee of the success of our undertaking. However this may be, we had been careful to see that all our paraphernalia had been put aboard the previous day, that there might be no unnecessary hurry at the last moment. On the morning of the day mentioned we ourselves went aboard fully prepared for our long journey. Our enthusiasm was undeniably at a high pitch, and it must be confessed that in our minds there existed no misgivings as to the final success of our mission. Promptly at nine o'clock all was ready, and the Company's tug towed the Gallileo from her berth out into mid-stream, and we were off on our trip to the other end of the world. But we were soon destined to a slight disappointment and delay. Hardly had we got under way than the clouds, which all the morning had hung low and threatening, settled down to the very surface of the water and buried the great city and the bay in a dense fog. This compelled us to come to anchor in the lower harbor, where we were detained for the remainder of the day and night. A little exasperated at this delay, we made the best of it, and together with the other passengers employed our time in writing letters to friends and relatives, to be despatched by our pilot when he should take leave of our vessel at the Scotland lightship.

All day and night we lay at anchor in this fog, while from fog horns, steam whistles and sirens there issued an uninterrupted and, to the landsman unintelligible, pandemonium of noises, not altogether unpleasing, however. I retired late at night and awoke the following morning to find, much to my relief as well as that of the other passengers, be it said, that the fog had cleared away during the night, and that while still cloudy, the atmosphere was clear with a stiff March wind blowing sufficiently strong to give promise of a decidedly choppy, if not unpleasantly rough sea outside the bay.

Our anchor was soon up and we steamed slowly down the channel, passing in succession all the familiar landmarks, until we arrived at the Scotland lightship, just outside Sandy Hook, where we dropped our pilot, having entrusted to him the letters written the previous evening.

We were now off on our long voyage, and I was glad that it was so, for already I began to feel that the undertaking for which I had been planning and looking forward, often under most discouraging circumstances, during the previous two years, was now in a fair way to be realized. The journey from New York to Buenos Aires did not differ materially from other similar sea voyages. There was perhaps a little more than the average of bad weather in the first two weeks of our trip, as was to be expected during the month of March. We encountered rough weather from the start and it continued thus for a week. Notwithstanding that Captain Braithwaite assured us from the first that it would change and for the better as soon as we were out of the Gulf Stream, the weather constantly grew worse, so that on the second morning I enjoyed the distinction of being the only passenger at the breakfast table, and the captain paid me the doubtful compliment of being a good sailor, remarking that I need never fear seasickness. He little knew how earnestly I was at that very moment striving against that very ailment. The weather continued unabated for a week, confining the passengers to the lower cabin and the ladies to their staterooms. When on the morning of the fifth day out, after the skylights had been broken in and the saloon and several of the cabins partially filled with water, much to our discomfort and the injury of our baggage, I casually remarked to Captain Braithwaite that I had not known that the Gulf Stream was so wide. He quite lost his temper for the moment at this bit of pleasantry on my part, and devoured with even greater avidity the plate of porridge which was all that remained from the third breakfast which had been successively cooked and washed overboard from the galley on that eventful morning.

After ·the first week the wind subsided, the temperature grew more congenial, and the sea calmed, so that the journey was henceforth a pleasant one, save for a two days' "pampero" off the coast of Brazil, when we were compelled to slow down with just speed enough on to keep under steerage way.

Our route lay direct from New York to the river Plate, and owing to the bad weather we were twenty-six days in making the trip from New York to Montevideo, which otherwise would have been accomplished in twenty-three days. Aside from these delays from rough weather, the voyage, while full of interest to myself from the novelty of never having been on a long sea journey, differed little, as I have already remarked, from other similar voyages. The sail through the tropics was not only interesting, but delightfully pleasant. I did not experience that discomfort from the heat that I had anticipated, though the inconveniences from this source were certainly more considerable on this than on any of my five subsequent voyages. This was doubtless due to the interior arrangements of the ship. For days we steamed through the bright green waters of the tropics, apparently on the borders of the Sargasso Sea. As I watched the brown masses of seaweed, *Sargassum bacciferum*, which lay on the surface, or was attracted by the peculiar beauty of the brilliant green of a bit of the same plant as it appeared floating a few feet or fathoms below, I could not but recall the renewed hope and encouragement its presence here had aroused in the hearts and minds of that Genoese navigator and his band of mariners who four hundred years before had embarked from a little Spanish port to sail these very seas in the successful quest of a new world, nor could I refrain from speculating as to whether the indomitable will of Columbus might not have been shaken and he persuaded to port his helm and retrace his course, if he had not mistakenly interpreted the presence of this plant as indicative of a not far distant coast. Who shall say how long America might have remained undiscovered, or what indirect influence this little plant has had on modern civilization, through the impetus given to the study of navigation and colonization by the discovery of the western hemisphere. Securing a bit of strong wire I con-

structed a large four-pointed grab-hook and then through the kindness of
one of the sailors I procured about fifty feet of small cord, which, when
attached to the former, made of it a very effective instrument for catching
isolated bunches of the Sargassum from the steamer's side, as we steamed
along.   Among the intricate branches of this plant live a great variety
of small crustacea, worms, etc., while the surface of the stems themselves
are frequently covered with the calcareous tubes of the more minute
annelids or encrusted with bryozoans of most delicate lacelike patterns.
But there are other things to interest and occupy the mind as one steams
along through this portion of the tropics.   From the surface of the water
rise numerous schools of flying fish.   It was interesting to note how they
would start suddenly from the water and continue their course, directed
upward at a considerable angle, until attaining a maximum altitude of
perhaps thirty feet, when instantly their course would be altered to that
of an incline, and they would descend until near the water's surface, over
and just above which they would glide, rising and falling with such perfect rhythm, as the crest and trough of each successive wave was encountered, as to maintain at all times a nearly equal elevation above the surface.
Then suddenly, when the original momentum had been expended, they
would plunge directly downward and immediately disappear beneath the
surface of the water.   The total distance covered at one flight would be
perhaps about three hundred yards, and in general appearance it was not
unlike that of the flight of a covey of our common quail, when startled
from a growing field of grain, with a breeze just sufficient to cause the
surface of the latter to move in gentle undulating waves.   All the individuals of each group rise almost instantly and take the same direction,
all settling in each instance in the same immediate vicinity.   The whole
flight is accomplished in such manner as to suggest, that in either instance
each animal pertained to an automaton, or at least, that one controlling
mind dominated and directed the action of each.   Moreover, the normal
distance of the flight of the quail does not differ materially from that of
the flying fish, though the former is without doubt capable of a much
more sustained flight than the latter.

Occasionally there could be seen from the side of the steamer a
"Portuguese Man of War," *Physalia atlantica*, gliding along upon the
surface of the water like a miniature ship with most beautifully colored
sails.   These and other objects served during the day to fill in the inter-

vals left over after the customary recreations aboard ship, while at night many otherwise tedious hours were very pleasantly passed in watching the phosphorescent light of the surface waters. This light is due to the myriads of animalculæ inhabiting the surface waters of this region to the depth of a few feet or fathoms. So abundant are these organisms in places, that with the least disturbance the waters become illuminated and during the night the crest of each wave appears as a broken line of beautiful white light. Over areas of more violent disturbance the illumination extends for some distance beneath the surface. About the prow and in the wake of the ship this light was often of such brilliancy as to illuminate the waters to a depth of several fathoms.

From the first two days after leaving New York we were entirely without the route of ocean-going steamers, and during the succeeding ten days our course lay through a trackless waste of sea, where not a sail appeared to bear us company. On the evening of the fifteenth day out a little excitement was exhibited among the passengers due to the announcement by Captain Braithwaite that at ten o'clock we should be abreast of Pernambuco, and should probably pass close enough in to see the lights of the town. As we had sighted no land since leaving New York, and this was to be the only land we were to have an opportunity of seeing until approaching Montevideo at the mouth of the River Plate, we were all to be found in the early evening pleasantly seated on deck, carefully scanning the horizon for the first indications of land. About nine o'clock the officer on watch came aft and reported to the captain a revolving light on our starboard bow. Not one of us had as yet detected it, though all had been intently watching for it, nor was it quite distinguishable to us until some time afterwards. Finally there appeared in the distance a dim light, which at first seemed to flash out at somewhat uncertain intervals. As the vessel proceeded on her course, the intensity of the light increased and the intervals between the flashes were seen to be quite regular. Later in the evening we came abreast of Pernambuco, although so far out at sea were we that it was impossible to distinguish either the merits or demerits of this Brazilian city. With the following morning we passed out of view of land, which we did not again sight until about noon of the 24th of March, when we raised the coast of Uruguay, and on the afternoon of the same day we passed between Lobos Island and the mainland. Early on the morning of the 25th we were abreast of Maldonado, and the gilded dome of its cathedral, apparently the only structure of any considerable dimensions in the town, was plainly visible among the other buildings of the village, which was partially buried in a somewhat scanty forest of palms, figs, orange, and other semi-tropical trees.   We were not long in reaching Montevideo, where we were promptly received by the master of the port and health officer. Here we were to stop for the day to discharge cargo.   We were only too glad to improve the opportunity thus offered of spending a day on shore. For however novel and interesting a sea voyage may be to the uninitiated, confinement for twenty-six days within the limited area of a steamer's deck is likely to create in one a desire for a wider sphere of action.   We were soon on shore and busily engaged regaling ourselves with the sights and pleasures of the city, all of which we thoroughly enjoyed, not so much from their distinctive merits as from their novelty.   In style of architecture Montevideo does not differ materially from other Spanish-American cities.   This has usually been described by most writers as exceedingly ugly.   To myself, however, there is something about those low buildings constructed of solid masonry, with their tiled roofs and broad and closed patios fitted with substantial if not always comfortable seats and decorated with a variety of tropical plants in fruit and flower, that is, to say the least, not entirely without beauty.   While there is in this architecture nothing of the grandeur to be found in that of either the Roman, Greek, or Gothic style, nevertheless these old buildings which have stood for several centuries are certainly even now much more becoming than that enormous structure known as the "College" built after the modern "dry goods box" style of architecture, which stands as the most conspicuous object in the foreground of the picture formed by the city as viewed from the deck of vessels lying in the roadstead.

Whatever the individual verdict as to the architectural beauty of Montevideo, all will, I think, agree with me as to its cleanliness.   During my two visits its streets struck me as being freer from dirt than those of any other city I had ever visited.   This is not because of the excellence of its municipal government as a whole, or the street cleaning department in particular.   It is due rather to the peculiarly advantageous topography which causes the drainage of all the streets to flow away from the city, the slope being so considerable and heavy showers so frequent, that the streets are at almost all times kept by nature free from filth.

After a most delightful day spent on shore we went aboard the steamer in the evening, and that night, under the direction of the company's pilot, proceeded one hundred and fifty miles farther up the River Plate to Buenos Aires, the capital of the Argentine Republic, the metropolis of South America, and a city which Argentinians would have us believe is the Paris of the New World.

Early on the morning of the 26th of March we anchored in the harbor of Buenos Aires, and here we had our first, but unfortunately not our last, experience with the dilatory methods of the Argentinians. We waited patiently hour after hour for the receiving and health officers. It was twelve o'clock before these put in an appearance, and even then we were somewhat fearful as to whether or not we should be allowed to enter, since, for some unaccountable reason, the company's agent at Montevideo had neglected, on our leaving that port the evening before, to return our New York health certificate. If the health officer saw fit, we could be detained until the latter was forwarded. The captain was in doubt regarding the better course to pursue in the matter, when the ship's physician (who, by the way, had been afflicted with seasickness throughout a considerable portion of the voyage), a recently graduated Yankee medical student, solved the problem by proposing that a bogus health certificate be substituted for the missing real one, since the inspector would doubtless not be able to detect the fraud. This proved an excellent plan, and upon the arrival of that officer, we were immediately received and allowed to enter. We steamed slowly into the berth assigned us in the "Boca," which might very appropriately be styled the "Erie Basin" of Buenos Aires. Here, upon the presentation of our credentials to the customs officers, we were very courteously treated and allowed to enter free all of our somewhat voluminous equipment, including fire-arms, ammunition, alcohol, and other articles ordinarily subject to a very high rate of duty. In each of my subsequent trips to Buenos Aires I was accorded the same courteous treatment at the hands of her customs officers.

We were not long in attending to our luggage, when, after bidding good-by to the captain, ship's officers, and our fellow-passengers, several of whom we were to meet afterwards under more pleasant circumstances, we hurried away, having engaged a hackman to drive us to the Hotel San Martin, situated in the midst of the business portion of the city, on a street of the same name and not far distant from the Plaza Victoria.

While the San Martin had been recommended to us, it must be confessed, that from an American's point of view the accommodations were exceedingly meagre and prices correspondingly high. Nevertheless, we made it our home for the nearly three weeks that we were compelled to remain in the city, while awaiting the departure of a steamer for Tierra del Fuego and the Atlantic ports of southern Patagonia.

On the following morning, March 27, we set about ascertaining the best means of reaching Gallegos, a small port at the mouth of a river of the same name, situated some thirty miles north of the eastern entrance to the straits of Magellan, in about S. lat. 51° 30′, which long before starting we had selected as the base of our operations while working in Patagonia. We called on Mr. Wm. I. Buchanan, at that time American minister to Argentina, and on the U. S. consul, Col. Edward L. Baker.[1] Both of these gentlemen were at once interested in our undertaking, and Col. Baker especially did much to aid us in our preparations. We soon learned that the Villarino, a small Argentine transport, was to sail from Buenos Aires for the south on April 16. This seemed a long wait, but since nothing better could be done, we had to make the best of it.

In order to travel on these government vessels, it was necessary to get certain permits from the proper officials. I had been apprised of this before leaving home by the Argentine minister at Washington, who had very kindly taken the precaution of writing me a letter of introduction to the Argentine Minister of the Exterior in Buenos Aires. Evidently our friend in Washington had done more for us than to write a mere formal letter of introduction, for when we called on the minister in Buenos Aires and presented this letter along with our other credentials, we were well received, and on taking leave were given an order for the free transportation on any of the government steamers of ourselves and our equipage from Buenos Aires to Gallegos and return. This was indeed making matters easy for us, and this thoughtful act of the Argentine minister no doubt saved us much vexatious trouble, for without his kindly aid we should doubtless have experienced great inconvenience in making the necessary arrangements for our departure, which would no doubt have been somewhat augmented through our lack of an adequate knowledge of the language and customs of the country. So obliging were our

[1] Since deceased.

Argentine friends that after a few days in Buenos Aires we were fully prepared for and anxious to commence our voyage to the south.

We were constantly urged by all our acquaintances, that, since winter was just setting in, we should do well to postpone our start until the following spring, some even declaring it impossible for us to withstand the rigors of a Patagonian winter.   However, we had tented it for many years on the wind-swept plains of Wyoming, Montana, and the Dakotas, often with the thermometer far below zero, and had no uneasiness as to our ability to survive successfully whatever blizzards Patagonia might have in store for us.

Having completed our arrangements for leaving Buenos Aires, we spent the remaining days left to us before actually starting on our trip southward in examining whatever of interest was to be found in and about the city, and in paying a visit to La Plata and its really splendid museum. Here we got a glimpse of that extinct mammalian fauna, in the remains of which we were soon to find the rocks of Patagonia so marvellously rich.

It was with no little pleasure that we welcomed the 16th of April, the day set for our sailing for the south coast.   While our stop in Buenos Aires had been pleasant, yet it had been more prolonged than we had anticipated, and we were really quite anxious to be again under way. We had been told the evening previous that the Villarino, the vessel which was to carry us south, would leave promptly at ten o'clock on the morning of the sixteenth, and we were especially cautioned to be on board early.   It is needless to say that this was scarcely necessary. However, at nine o'clock of the morning in question we were on board. Then began a tedious wait of several hours, so that it was late in the afternoon before our little vessel cast loose and steamed slowly out of the "Boca" on her way to the south coast of Patagonia and the Fuegian ports.   What a feeling of relief it was to be on our way once more. Slowly we steamed out along the channel into the River Plate.   The day was an ideal one, even for Buenos Aires, where so many days are ideal. The mild warmth of the sun softened the temperature of a crisp autumn breeze, which scarcely sufficed to produce a ripple on the surface of the river.   The latter appeared like a great mirror, or, as its name implies, a mighty stream of silver, moving slowly seaward.   Every molecule throughout its sixty miles of breadth acted with such unison, that from its bosom there was reflected an exact image of the shipping at anchor in the roadstead, or of the not too distant shores, as from a polished surface of silver. A noble stream indeed is the River Plate, when from this point, at a distance of 150 miles from its mouth, its breadth is such that it is impossible to see the opposite shores of Uruguay, nothwithstanding the advantageous atmospheric conditions. The favorable impression which this river first made upon me was enhanced rather than decreased by a journey made some three years later, for a distance of more than 900 miles up its waters into the interior of South America. As the sun disappeared beneath the western horizon the last of the cathedral spires of the great city were lost to view. With fine weather and a smooth sea we made good progress, and on the morning of the 18th of April, when I awoke and went on deck, a most novel sight was presented. All about the ship's sides were great salt marshes covered with a luxuriant growth of grass that stretched away for miles on either hand. Straight behind and in front of us was a narrow channel of water, so narrow indeed, that as we looked about, the impression received was that of steaming through a great morass or swamp, a rather unusual procedure for an ocean-going vessel. We were in fact entering the port of Bahia Blanca, styled by the Argentinians "the Liverpool of the South," though I thought this a somewhat too aspiring title. We were soon at anchor near the end of the only mole of which at that time the "Liverpool of the South" could boast. At this mole three or four large freight steamers were busy loading wheat and cattle for Europe, which were delivered at the mole by a single track railroad, that connected the port with the city of Bahia Blanca, located some miles inland.

Since we were to spend the greater portion of the day here in embarking a company of some sixty Argentine soldiers to be taken to Santa Cruz, we improved the first opportunity of going on shore. From a scenic standpoint nothing could be more uninteresting than the country immediately surrounding this port. The almost level surface seemed but recently to have been recovered from the sea and was but poorly covered with vegetation, consisting for the most part of low grasses, with occasional shrubs and bushes. Even this scanty vegetation occurred only in patches, such areas alternating with others of equal or even greater extent, where, instead of a scanty vegetation, the surface was covered with alkaline salts (chiefly sodium sulphate) which in dry weather has the appearance of a light fall of snow. Away in the distance, forty miles to the northward, Sierra Ventana raised its hoary head of gray quartzite, sentinel like, 3,000 feet above the surface of the surrounding plain, and formed the only interruption upon the level surface of the seemingly limitless expanse of country, that stretched far as the eye could see to the south, west and north, its dull color due to the scanty covering of brown withered grass, otherwise entirely unrelieved.

Time did not permit our visiting Punta Alta, where Darwin had a half century before discovered the remains of Megatherium, Megalonyx, Scelidotherium, Mylodon, Toxodon, and a host of other equally interesting and long since extinct animals. Everywhere in the immediate vicinity of the port the beach consisted of mud flats. Some two miles inward, however, workmen were engaged in digging a large well for the railroad company and from a bed of marl some eight feet below the surface I secured a considerable number of shells belonging to several species, apparently not differing from those now found in the waters of the bay.

Early in the afternoon the warning whistle blew for us to come on board, and we immediately repaired to the mole, where we found that the Villarino had taken her position alongside one of the larger freight steamers. We were soon aboard, and as the company of soldiers before alluded to, together with their officers, had already embarked, we were not long in getting under way and steaming out of the port of Bahia Blanca by the long narrow canal by which we had entered it. The company of soldiers just mentioned was composed of about sixty men, and a dirtier, more shiftless and less sightly aggregation of human beings it has never been my lot to see. Their clothing, luggage, and persons were all in a filthy condition. I am speaking now of the rank and file only, for in justice to the officers it must be said that their appearance was always quite the opposite from that of the men, and I never could understand why these officers were so lax in enforcing discipline and cleanliness in the ranks. Whatever the cause, the Argentine soldiers and sailors, when compared with the standard maintained in our army and navy, are painfully lacking in that prepossessing appearance which characterizes our men, by reason of their neat, well-fitted uniforms, tidy appearance and erect military bearing.

Our next stop was at San Blas, the name given to a small land-locked bay, some sixty miles north of the mouth of the Colorado River. The beach here is composed entirely of shingle, and this was our first view of that great shingle formation which is everywhere found covering the Patagonian plains to a depth of from a few to more than one hundred feet. The great extent and uniform nature of this has been remarked by many and has not yet been satisfactorily explained.

From the beach a low ridge rises to a height of perhaps thirty feet. On ascending this there appeared at a distance of perhaps a half mile a substantial farm house and outbuildings belonging to a gentleman whose acquaintance we had made in Buenos Aires. As we approached the *estancia*, we were surprised to meet this gentleman himself, who at once recognized us and made us welcome. We remained at San Blas for the rest of the day, most of which I spent in studying the surface of the surrounding country, which seemed to be made up of a succession of low parallel ridges, separated one from the other by only a few rods, in most instances, and with no great difference in altitude. These ridges I decided were the remnants of old beaches thrown up and left by the sea during the slow process of elevation by which the land has been brought above sea level. Late in the evening we went aboard the Villarino, and a little later we were joined by an Italian engineer employed by the Argentine government to inspect the different harbors of the coast, in order to determine which of them would be most suitable as a naval port for Argentina. Together with our host he had preceded our arrival at San Blas by a day or two, having come down from Buenos Aires in a small government steamer that lay at anchor in the bay. Our friend, who owned all the land in the immediate vicinity of the harbor, was very profuse in proclaiming to us the special advantages offered by San Blas as the seat of a great naval port, and had no doubt, with characteristic Irish shrewdness, employed much of the time since his arrival here in pointing out to the aforementioned engineer these highly praised, though to my mind somewhat doubtful advantages. After waiting for the tide to rise, late in the evening our anchor was hoisted and the little steamer moved slowly toward the harbor's mouth. As we were passing through the narrow channel at the entrance, there was first a slight jar, immediately followed by an almost deafening noise, caused by the grating of the ship's keel on the coarse shingle at the bottom. Almost immediately we were brought to a standstill at the very entrance of the harbor, which, we had just been told, possessed so many and such excellent advantages for the establishment of a great naval port.   If we, drawing only eleven or twelve feet, had to await a favorable tide, and then become grounded, what of the great cruisers and battleships that in future were to enter here to be drydocked and repaired ?   Fortunately the tide was still making, and an hour later we were just able to pass out and proceed on our way.

Our next stop was at Port Madryn in New Bay, the port of entry for the Welsh colonies on the Chubut River.   We arrived at the entrance to New Bay early one morning, and as we approached, a line of high cliffs rose sheer from the water to a height of perhaps 200 feet and stretched away on either side as far as the eye could reach.   This was our first view of that great sea wall that extends almost uninterruptedly all along the eastern coast of Patagonia, from the mouth of the Rio Negro to the eastern entrance of the Straits of Magellan, and with which we were shortly to become so familiar, through the remarkable advantages it presents as a collecting ground for both vertebrate and invertebrate fossils, certain species of recent birds and mammals and other objects of natural history.

As we entered the bay, which at its mouth has a breadth of hardly more than a mile, it was seen to expand rapidly into a broad, nearly circular basin with a diameter of little, if any, less than forty miles.   On this morning it was a particularly beautiful sheet of water.   Seen under the perfect atmospheric conditions that then prevailed, it was indeed a thing of beauty.   From its surface, smooth as that of a polished table, were reflected images of the high, precipitous cliffs surrounding it.   The chalk-like whiteness of the latter was intensified by the sunlight as it came streaming through the clear autumnal atmosphere unobscured by a single cloud.   The peculiar beauty of this body of water was emphasized by the dreary, not to say desolate appearance of the surrounding plains, occasional glimpses of which could be caught from the vessel's side, as we steamed along toward Port Madryn, revealing the bleak, treeless nature of its surface which only supported a scanty covering of brown and withered grass.

We soon reached Port Madryn, for which we had several passengers and considerable cargo.   Here we were detained most of the day discharging the latter.   Mr. Peterson and myself were on shore at the first opportunity.   Life in the waters of this bay must be extremely abundant and quite varied.   The beach is literally covered with the cast up shells of mollusks, both bivalves and gastropods.   Sea urchins and sand-dollars are also abundant, while sponges, hydroids and bryozoans may be seen everywhere in the greatest profusion. Protruding from the surface of a bluish sandstone underlying the shingle of the beach and from the face of the cliffs above were numerous shells of a large oyster nearly a foot in diameter, while Cucullea, Venus, Mytilus, Natica, Trophon, Pecten, Dentalium, Brachiopods, Scutella, and a host of other forms of invertebrates were found associated in the same rock. Late in the evening we left Port Madryn and steamed down until near the entrance of the bay, where in a little cove under a precipitous cliff we anchored for the night. I have yet to learn why we anchored here with such extremely favorable weather conditions, unless it was that Captain Calderon, who had at all times been very obliging, wished to give us an opportunity of seeing the splendid entrance to this magnificent bay under the mellow light of a full moon, with the Southern Cross adding its beauty to the general interest of all about us. Whatever the cause may have been, we anchored as detailed above, and, dinner over, the boats were lowered and manned, each with six mariners from our crew, and at the invitation of the captain Mr. Peterson and myself accompanied him and several of the ship's officers on a fishing expedition. We were well supplied with tackle and bait, and after several hours passed most pleasantly we returned to the ship, shortly after midnight, with a bountiful supply of most excellent fish, purchased from some fishermen at the "Pescadores," a few miles up the bay. Many were the expressions of disgust from our Argentine friends that evening because the fish would not bite, but to myself it mattered little, for I was chiefly interested in the splendid, if somewhat solitary, beauty of our surroundings, and was overcome and enraptured by the quite unexpected novelty of the situation. From childhood I had thought of the coast of Patagonia as visited by almost perpetual storms, surpassed in their frequency and violence only by the region about Cape Horn. But we had now been steaming along this coast for several days under almost ideal atmospheric and weather conditions. How different our actual experiences from the impressions given by reading Dana's "Two Years Before the Mast," or "Gold Diggings at Cape Horn," by John R. Spears. With what feelings of comfort and absolute safety we moved about from one locality to another over the perfectly placid surface of the bay, its waters undisturbed by even so much as a zephyr. The complete quiet, save for the rhythmic murmur of the muffled oars, was comparable only with that described by Arctic travelers.   This quiet suffered occasional interruption, as our boats moved slowly under the shadow of some overhanging cliff and suddenly startled, by our near approach, a group of water-fowl resting quietly at the base.   These would move off in great confusion, uttering at the same time those indescribable screams, which, for volume and harsh ness, are unsurpassed in other birds.   Thus, for a few moments, would the almost absolute quiet be changed to the greatest pandemonium. Then, as the disturbers gained a spot of fancied or real security, the noise would gradually die away, followed usually by the plaintive hoot of an owl, sitting solitary in his eyrie high up in the overhanging cliff.

A little after midnight we returned to the ship and were soon in our berths and fast asleep, after one of the most pleasant and interesting days of the entire trip.   Long before we had arisen on the following morning the anchor was up and we were off again on our way south.   When we came on deck in the morning, we were greeted with the same clear, bracing atmosphere, warmed by the sun's gentle rays.   Captain Calderon and several of the ship's officers, however, were very careful to tell us that we should soon encounter bad weather.   On asking if the barom eter indicated a change for the worse, we learned that their predictions were not based on any change of atmospheric conditions indicated by that instrument, and consequently took their predictions for what they were worth.   Notwithstanding their assurances that we should never cross the Gulf of St. George, which we were now entering, without encounter ing a storm, we resolved to make the best of the present, regardless of what the future might bring us.   For three days we kept on our way southward, all the time out of sight of land until approaching the entrance of Port Desire, at the mouth of the Desire River, and through out the weather was all that could be wished.

At Port Desire we anchored in front of some old Spanish ruins, which stood at a distance of about two hundred yards from the water's edge, on the level surface of a narrow terrace, at the foot of an escarpment of hard, coarse-grained brown sandstones.   The sandstone was overlaid by "Tosca," which formed the uppermost portion of the bluff, from the sum mit of which, stretching far away into the interior, lay the broad level plains so characteristic of Patagonian landscapes.   The ruins just referred to were constructed of a good quality of sandstone.   The foundations were of such a nature as to suggest strength and solidity, and the masonry was of a good quality. These ruins when last seen by myself in May, 1899, compared very well with the description given of them by Darwin, who had visited them sixty years earlier. Growing about were a few dwarf cherry, plum and quince trees, or bushes, suggesting that those who founded the settlement had faith in the hospitable nature of the surrounding country, and had not anticipated its early, if not almost immediate abandonment. From the nature of the ruins it is evident that, of the structure contemplated, only the foundations were completed. The harbor in front of the ruins is rather picturesque, more especially at low tide, when in its upper stretches it is dotted over by numerous small rocky islets, for the most part entirely submerged at high tide. The maximum difference between high and low tide in this port is twenty-eight feet. On the north shore of the port a series of rugged cliffs of red porphyry rise, in most places, directly from the water. On the south there is a low, level plain with some very picturesque chimney rocks, standing like so many giant obelisks on the surface, at a distance of perhaps a mile from the shore. The upper bay at all times, and especially at low tide, literally swarms with animal life. The recently submerged islands are exceedingly favorable feeding grounds for the numerous and varied water fowl that frequent this coast. Myriads of these birds may be seen everywhere, and at times they fairly cover the surface of the smaller islands. Engaged in a continuous struggle to procure, and defend when found, each choice morsel left stranded by the receding waters, and accompanying their frantic efforts with a series of distracting shrieks and screams, they present an extremely animated picture as they swarm in the waters and over and upon the islands.

We spent most of the day at Port Desire. Since we had been told that we were to replenish our supply of fresh meat here, those who had volunteered us this information, at the same time expatiating on the splendid opportunity we should have for witnessing the wonderful dexterity of the Argentine gaucho with the lasso, we were not a little curious to see with what neatness and despatch these lauded gauchos could do their work. This curiosity was no doubt enhanced by a certain familiarity with the expeditious manner in which such matters are usually attended to by the experienced cowboys of our western plains. What was our surprise at seeing a small bunch of cattle driven from the plain down near the beach, two of which were selected and, after many vain attempts, finally caught.   But that which was more striking even than their lack of skill with the lasso was their wanton cruelty.   When an animal was caught, instead of despatching it at once, its hamstrings were cut and it was left in this helpless condition to be jostled and trampled by the rest of the bunch during the half hour or more consumed in catching the second animal, which was served likewise.   When both were despatched and dressed, they proved to be old cows in very poor condition and in calf, one carrying a half grown fœtus, and the other, one about the size of a rabbit.   The larger of these was carefully dressed and taken on board together with the two adult carcasses, and was no doubt served up to us as soup during the remainder of our voyage.   Such a thing as unripe veal is apparently unknown among Argentine epicures.

Late in the afternoon we left Port Desire for Santa Cruz, where we arrived on the evening of the following day.   This settlement, which for years was used as a penal colony by the Argentine government, is situated on the south side of the Santa Cruz River, some eight miles from its mouth, in 50° 10′ S. Lat.   It was at this place we were to disembark the company of soldiers we had brought from Bahia Blanca, and since we also had a considerable amount of cargo to discharge, we were delayed most of the following day.   The village consists of some twenty or thirty small shanties constructed of adobe or corrugated iron, while at a distance of some two miles further up the river are located the buildings formerly occupied by the prisoners and their guards, when Santa Cruz was a penal colony.   These are now collectively spoken of as "The Mission," and are occupied by the master of the port and his "mariners."   The post-office for Santa Cruz is also located here, so that persons residing in Santa Cruz proper have to go two miles to get their mail.

On the morning following our arrival at Santa Cruz we went ashore. At low tide there is a great bank of shingle rising from the water to a height of some thirty feet.   At high tide this is for the most part sub-merged.   Extending back from the river for a distance of about a mile, is a nearly level valley, in which stand the few small buildings of the village.   This valley gradually increases in elevation, as one proceeds from the bed of shingle that forms the river's bank to the foot of an escarp-ment, which rises some three hundred feet above the surface and forms the southern border of the valley.   This small valley, like many another in this region, is of recent origin, and has been formed by the gradual elevation of the land. The shingle with which its surface is everywhere covered has been accumulated in the same manner as the bed just mentioned as occurring on the beach. Immediately in front of the beach there is a strip of perfectly bare, clean shingle, averaging some two hundred feet in width, its surface thrown up into a number of parallel undulating ridges or drifts. As one proceeds inland from the beach, these ridges gradually become less perfectly defined, the completely barren, shingle-covered surface becomes, successively covered over more and more with a scanty soil, resulting finally in a surface, which though but poorly clothed with vegetation may be considered as fairly representative of that of the Patagonian plains of this region. During our hurried walk across this narrow valley we became acquainted with most of the plants common to the plains of southern Patagonia, which were to be seen at that season of the year. Shooting up from the scanty soil, held between the small rounded stones, were scattered bunches of brown withered grass. On one hand there might be seen dark green patches of the "Mata verde," *Lepidophyllum cupressiforme*, intermingled perhaps with the taller and coarser stems of the "Mata negra," *Verbena caroo*, two of the commonest shrubs of Patagonia, usually attaining to a height of only two or three feet. Occasional thickets of the "calafate," *Berberis cuneata*, and "incense bush," *Schinus dependens*, at times growing to the height of eight or ten feet, were scattered over the surface of the valley. Concealed within the shelter of the branches of such thickets were usually to be found one or more skeletons of the guanaco, in varying degrees of disintegration, according to the length of time that had elapsed since, weakened by old age, disease, or the rigors of an unusually long and hard winter, they sought protection from some merciless blizzard, only to find death in the too scanty shelter offered by these bushes.

The delightful weather we had experienced while en route from Buenos Aires to Port Desire did not continue with us to Santa Cruz. While the wind was not sufficiently strong to render it uncomfortable on board ship, the atmosphere was damp and cold and the wind raw and disagreeable. There was little rain, but the skies were continually overcast with low, thick, leaden-colored clouds which seemed to shut out whatever heat the sun's rays may have had in this latitude at this season of the year.

On the morning of the second day after arriving at Santa Cruz we steamed out to the mouth of the river and anchored just off Direction

Hill, on the south side of the river, directly in front of a precipitous cliff, which rises to a height of three hundred and fifty feet above the water's level.   This was the identical cliff mentioned by Darwin and from which he made his section of the Patagonian beds, published in his "Geology of South America."   As I sat on the deck of our little steamer, with this volume before me, and carefully compared his description and section with the actual one as it stood plainly visible, I was struck with the remarkable lucidity and accuracy of that great mind, and wondered whether it would be possible for me to add anything of importance to the observations made and materials collected by him.   Each succeeding layer of light or dark colored sandstone, or clay was plainly discernible in the exact sequence detailed by Darwin.   Moreover, on going on shore an hour or two later with Captain Calderon, we found the different strata to contain the same fossils mentioned by Darwin.   For three long days we lay here at the mouth of the Santa Cruz River, under the pretense of erecting a signal tower on the north shore, which for several miles inland is low and flat.   Each day officers and men were sent ashore for the purpose of prosecuting this work.   Each day we accompanied them, and each day they returned, having neither accomplished nor attempted anything.   It was a weary wait for us, as the weather was all the time disagreeable.   We had now been nearly two months on our journey, and here we were at last within ninety miles of our destination, burning with anxiety to be at work, but detained day after day for no apparent reason.   Finally, shortly after midnight on the morning of our fourth day, I was awakened by the familiar noise of the ship's engines and realized much to my satisfaction that we were once more under way and headed for Gallegos. We were up and dressed bright and early in the morning and as soon as we came on deck, the same weather conditions greeted us as had prevailed ever since leaving Port Desire, save that the temperature had fallen somewhat and the wind had increased considerably in force.   We were scarcely on deck, when on looking about us, we detected the backs of some huge black objects in the water, and suddenly discovered that we were right in the midst of a school of whales, probably *Balæna australis*, for this species is known to frequent this coast and to be gregarious.   I counted no less than fourteen of these great monsters as they disported in the water about the ship.   Frequently they would come to the surface and glide along with their great dorsal regions protruding from the water, while occasionally one would be seen to elevate his large forked tail and several yards of the posterior portion of the body high in the air, then, plunging suddenly and apparently almost vertically downward, immediately disappear beneath the surface. They seemed to take no alarm at our presence among them, appearing rather to enjoy our company, for I noticed that for some moments several of them took a position just forward of the vessel's bow, where they disported themselves in much the same manner as I have on other occasions observed porpoises to do, crossing and re-crossing the vessel's course with the greatest ease, while at the same time maintaining the same forward motion as the ship. I had not supposed that these whales could swim so rapidly. The excitement and amusement caused by their presence was soon terminated, for they shortly disappeared almost as suddenly and quite as mysteriously as they had appeared.

After this pleasant distraction, all too quickly passed, we had time to look about us and discern if possible the entrance to the Port of Gallegos. Far away in the distance could be seen a distinct, but perfectly straight and unbroken line resting against the western horizon. To our inexperienced eyes it was not possible to say definitely if it were land, for what must be the nature of a country presenting such an apparently continuous and unbroken shore line as this? As we approached more nearly the mouth of the river, our conjecture was verified, and as Cape Fairweather came in view on the northern shore, a rugged line of perpendicular cliffs, four hundred to five hundred feet in height, was seen to extend to the northward as far as the eye could reach. The flat, level plain above terminates abruptly in the escarpment forming this sea wall. The perfectly level surface of this plain is absolutely unrelieved by an elevation, even of a few feet.

Turning to the southern shore, to which we had now approached quite near, it was seen to consist of a low, level plain, only slightly raised above the water's level. In the foreground was a broad, shingle-covered beach on which with the aid of our field glasses we could distinguish a small band of guanaco, the first of these animals we had seen. A few miles distant in the interior were a number of black, rugged piles of basalt and other volcanic materials, remnants of volcanoes from which in no very remote times, geologically speaking, the streams of lava now covering the plains at their bases poured forth. These extinct volcanoes, known as the

Friars, rise several hundred feet above the surrounding plain and are but examples of others that occur in considerable abundance throughout the Patagonian plains.

We arrived at Gallegos, the seat of government of the Territory of Santa Cruz, at about ten o'clock on the morning of April 30, just two months after leaving New York.   This port is situated on the south side of the Gallegos River and about seven miles from its mouth.   We had come in with a rising tide and anchored just in front of the small village, which at that time consisted of not more than twenty-five houses, for the most part small and miserably built of wood and galvanized iron, carelessly thrown together.   After we had anchored, I was utterly amazed to see with what rapidity the incoming tide rushed by the sides of the vessel. The current was like that of a mill race, with a velocity of from six to seven miles per hour.

The morning had been cloudy, and a fine mist of sleet and rain had set in, driven by a stiff southwesterly breeze, which intensified the already disagreeable nature of the weather.   Shortly after coming to anchor, a boat was lowered and manned, and Captain Calderon, an Italian marquis who had been a passenger from Buenos Aires, Mr. Peterson and myself went on shore and proceeded at once to the government building to pay our respects to the governor, General Edelmiro Mayer, by whom we were very cordially received, after being formally introduced by Captain Calderon.   Upon presenting to the General our credentials and the various personal letters of introduction with which we were supplied, we were invited to be the guests of himself and staff during such time as we should necessarily be delayed in getting together the horses and other equipment for our work.   We were only too glad to accept this invitation, since there was not a comfortable hotel or lodging house in the place.

CHAPTER II.

In Gallegos: Courteously received by Governor Mayer and Staff: Fire as a means of comfort unknown in South America: Discomforts of Gallegos: Enormous rise and fall of tide in river at Gallegos: Antiquated methods of overland travel and transportation of produce in Patagonia: Visit to Governor Mayer's estancia: Guer Aike: Mata verde: The plains of Patagonia: Laguna Leona: A diminutive oasis due to a small spring of sweet water: its effect upon the fauna and flora: Origin of Laguna Leona: Coy River: Two classes of rivers in the Patagonian plains: Visit to an Indian encampment: Mate drinking among the Indians and Argentinians: We return to Gallegos: Purchase of horses and cart: Peculiar custom of pulling horses from the cinch in Argentina: Leave Gallegos equipped for work, May 16, 1896.

THE walk from the beach to the government building had practically been a wade through the mud, so that we all arrived in a somewhat uncomfortable condition. The building was a large, barn-like structure two stories high, built of native lumber. Without, it was weather-boarded and roofed with galvanized iron. Within, it was unplastered, but sealed with rough lumber covered with cheap cloth and then indifferently papered. In one room there was a billiard table, and the Governor's private library of some six thousand well selected volumes, consisting about equally of French, Spanish, German and English, for the General was an accomplished linguist. In another room stood a grand piano and a handsome American organ. Notwithstanding all these articles which in our own country might be taken as indicative of a comfortable residence, there was naught of real comfort about the place. Although we arrived in a condition which at once bespoke our discomfort, and were received with the greatest cordiality, nothing was done to relieve our distress, due to wet feet and water-soaked shoes. We were ushered into no reception room provided with easy chairs and made comfortable with the glowing warmth of a cheerful fire. To use a common expression, the house was cold as a barn and discomfort was plainly depicted on the countenance of every one present during the two hours we alternately sat and stood, while awaiting the preparation of the breakfast to which we had been invited. It was with great anxiety that I awaited the announcement of this meal, not that I was suffering from hunger, but in the vain belief that the dining room would at least be kept sufficiently warm to enable us to eat in comfort. What was my disappointment, on our being at last ushered in to breakfast, to see a long table elegantly laid, set in the middle of a great room, which, if there was any difference, was even colder and more cheerless than that we had just left. However, with Mrs. Mayer as hostess, we were soon enjoying a splendid breakfast. Everything on the table bore evidence of having been selected and prepared by no inexperienced hand. Our host, who from the very first moment of our acquaintance had kept up an incessant conversation, asking of us all manner of questions about the States, where, during his youth, he had spent four years as an officer in our civil war, evinced the keenest interest in our undertaking and was even profuse in offering us assistance in every way possible. There were at breakfast a German-, an Italian-, two English- and four Spanish-speaking people, and it was remarkable to see with what rapidity and fluency our host carried on a conversation with all, addressing each in his native tongue and turning from one to the other with the greatest ease.

Breakfast finished, we amused ourselves for a time in the library, which, while generally well selected, contained a few volumes highly prized by the Governor. One such in which he evidently took great pride was a superb copy of Poe's "Raven," printed in large folio and profusely illustrated with large hand-colored plates. While in America, General Mayer became a great admirer of Edgar Allen Poe, and after returning to Argentina, translated his works into Spanish.

All this and much else that the General had to show and tell us was quite interesting, but try as we would we could not forget our present discomforts, due to the feeling of dampness and cold that pervaded everything about the house. It was not that our host was unmindful of our welfare, for it was quite apparent that he was not only willing but anxious to be of service to us. Fire or other artificial heat as a means of comfort was until very recently quite unknown in South America, and even now it is exceptional even in the palatial homes of the wealthier Argentinians in Buenos Aires.   Accustomed all his life to these conditions, it never occurred to him that a fire would add very materially to our comfort, and little good would it have done if it had, for there were no conveniences for making a fire.   It will doubtless seem strange to those accustomed to all the many comforts of our homes, built and fitted with all modern conveniences, to be told that here in south latitude 51° 30′, or six hundred and fifty miles farther from the equator than is New York, in the principal government building of the capital of a territory with an area equal to one half that of the German Empire, there is no provision for heating either the entire building or any portion of it.   Yet such was the fact.

It was with considerable pleasure that late in the afternoon we received the information that our luggage had arrived from the steamer.   We were immediately shown our quarters and proceeded forthwith to provide ourselves with heavier and more comfortable clothing, with which, fortunately, we were well supplied.   Clad in dry and warmer clothing we were, if not what might be called comfortable, decidedly less uncomfortable than we had been during the day, and after a splendid dinner followed by some delightful music, with the Governor and his wife as the performers, we retired for the night, fully realizing that our stay in Patagonia was not going to prove a pleasure trip from the standpoint of the summer tourist, however successful it might be from a purely scientific point of view.   We were determined, however, and felt confident of our ultimate success.   While inwardly disgusted with the mode of life of the natives, which had in it so few elements of real comfort, yet we resigned ourselves to it with all the grace possible during the short time it was necessary for us to remain in Gallegos, while getting together our outfit, purchasing horses, etc.   We well knew that when once we were ready to start and dependent upon our own resources, we could, with our equipment and experience in camp life, make ourselves far more comfortable in our tent than here in these cold, damp, cheerless halls.   Moreover, we should then be under the exhilarating influence of work to be accomplished, with an abundance of most interesting material about us on every hand to be had for the taking.   It was with some such feelings as the above that we retired for the night, our first to be spent on shore in Patagonia.   While we earnestly endeavored to adapt ourselves to the customs of the country, in some respects this was quite impossible.   All over South America two meals a day are considered sufficient, so that breakfast and dinner are the daily allowance. The hours for the former are from eleven A. M. to one P. M.; those for the latter from five to eight P. M. Strive as we might, it was quite impossible to restrain our appetites until the breakfast hour, which with our hosts was always set for one o'clock. We had always been accustomed to rising at a seasonable hour in the morning, and refreshing ourselves with a substantial meal before essaying any work of more than minor importance, and to wait until one o'clock for the arrival of this accustomed meal, was an imposition against which our appetites seldom failed to rebel, whenever and wherever put to the test, so long as we remained in the country.

We were up early the following morning and were delighted to find that the weather, while still raw and cold, showed a decided improvement over the day previous. On looking toward the river, I was surprised to see only the top-masts of the Villarino showing above the crest of the shingle-covered beach. On entering the government building a little before noon, after leaving the vessel the day before, I remembered having seen almost her entire hull, as she lay tugging at her anchor on the broad bosom of the river. Could it be that some accident had befallen her during the night? The entire village was still fast asleep, so that there was no one at hand of whom to inquire as to what had happened. Just as I was becoming thoroughly alarmed for the safety of the vessel and those on board, I noticed a great change in the character of the river itself from that which it had presented before. Instead of the broad sheet of water stretching uninterruptedly for three miles, almost on a level with the surface of the uppermost banks of shingle on either shore, there was now only a narrow channel to be seen on the northern side, while an extensive mud flat lay between it and the main river channel on the south, which from my point of view was quite concealed by the bank of shingle in the foreground. It was in the latter channel that the Villarino was anchored. This was my first practical illustration of the enormous tides in this river, which are only exceeded by those of the Bay of Fundy and are given by some authorities as attaining a maximum of fifty-two feet.

We spent the first day after our arrival at Gallegos in overhauling our luggage, in order to ascertain what, if any, injury our equipment had suffered during the long voyage. We were delighted to find that everything was in excellent condition.

We immediately commenced making inquiries as to the best means of procuring horses and a suitable vehicle with which to get about over the country. We had unfortunately relied upon the advice given us by Mr. John R. Spears, who had a year or two before visited the country, going down and back on the same steamer, making only the short stops of the vessel at each port, and from the meagre information thus hastily gained returned to write his "Gold Diggings of Cape Horn," a quite readable but very unreliable book. Mr. Spears had given us, before sailing from New York, the assurance that we could procure in Gallegos American mountain wagons, harness, and everything needful for a trip inland. What was our disappointment on our arrival to find that the only vehicles to be had were cumbersome two-wheeled bullock carts, heavier by half than any load which they might be trusted to carry. Some of these were constructed in Buenos Aires, while others were of English manufacture. But, whatever their manufacture, they were always excessively heavy and clumsily made, no thought of combining strength with lightness having apparently ever entered the minds of those by whom they were built. Nothing more stupid and ridiculous can be conceived than the present method of freighting overland the wool and camp supplies of the Patagonian sheep farmers. The axle of these carts usually consists of a steel casting, seven feet long and three inches square in cross-section in the middle, with either end rudely fashioned into a spindle for the wheels, each of which will, as a rule, weigh from three to four hundred pounds. The body is framed of four- by six-inch timbers and is usually about eight feet in length by five in breadth. It is provided with a floor and sides, but is without ends. The tongue or pole is usually made of a great beam six by eight inches in cross-section and eighteen feet in length. It extends underneath the entire body of the cart and is firmly bolted to it and the axle. When set up, such a cart will weigh about three thousand five hundred pounds, while the maximum load for one cart, with two men and from four to six bullocks, was seven bales of wool weighing some four hundred pounds each. It is difficult to understand why in a country like Patagonia, naturally so well adapted for overland travel, the inhabitants, for the most part either Scotch, English, or German, are so backward in introducing more improved methods for the transportation of their wool and necessary ranch supplies to and from the different ports along the coast.

Not only were all the vehicles unnecessarily heavy and cumbersome, but it was difficult to find any that were for sale. An enterprising Chileno with a more enterprising German wife, Señor y Senora Cayetano Sanchez, were kind enough to offer us an ancient specimen with wheels about two and one-half feet in height for the modest sum of three hundred and eighty dollars, which they assured us was *muy barato* (very cheap). And it looked indeed for several days as though we should be compelled to accept this vehicle, notwithstanding its generally dilapidated appearance and low wheels, which would doubtless have been the cause of no little inconvenience in crossing the various streams. Since we had not yet secured any horses, we were in no immediate need of a vehicle, and hence postponed our negotiations with Señor Sanchez, hoping in the meantime, like Mr. Micawber, that something would turn up to relieve this particular phase of a really embarrassing situation.

We were told daily by Governor Mayer and Messrs. Aubone and Villegrand that we should have no trouble in getting all the horses we should need, and after a few days we were fairly besieged with parties offering us all sorts of nondescript horses at ridiculously high prices. We had been invited by the Governor to make a trip to his estancia, located some sixty miles northwest of Gallegos, on Coy River. We were of course only too glad to avail ourselves of this opportunity for a trip inland, in order to see for ourselves something of the nature of the country, so as the better to judge of the requirements necessary for the successful prosecution of our work. We were to accompany Señor Villegrand, who, with two policemen and a young Argentinian, was to take a troop of some sixty or seventy young horses from Gallegos to the estancia.

Late on a Saturday evening Señor Villegrand informed us that we should start the following Sunday morning, at the same time asking me if we had rather ride horseback or in a carriage. When I told him that we should much prefer to go on horseback, he expressed some doubt as to our being able to make the journey. We assured him, however, that sixty miles in a day was not considered an unusual ride on horseback in our own country, and since he did not intend making the entire distance in one day, we had little doubt that we should be quite equal to the undertaking. The next morning we were up bright and early, anxious to start on what was to be our first trip over the Patagonian pampas. We had been told that we should start "*muy temprano*" (very early), which we were soon to learn meant in Argentina any time after eleven A. M. After we had waited for several hours, Señor Villegrand made his appearance, our horses were brought out and saddled with our own saddles, while the Señor with the assistance of pretty much the entire official staff, including the Governor and the police force, succeeded in harnessing and hitching to a heavy, cumbersome, two-wheeled English cart, fitted with shafts, a rather stylish looking but ridiculously small pony, considering the size of the cart. When, after a considerable delay, all was in readiness for the start, this cart-horse, notwithstanding the many virtues of which he had been said to be possessed, proved to be quite refractory, continually insisting on going backward rather than forward. However, after some delay, with a little encouragement he was induced to make a start, and we were soon on our way to Guer Aike, eighteen miles up the river from Gallegos, where we were to stop for the night. The young Argentinian and the policemen had gone on ahead with the troop of horses, while Mr. Peterson and I remained behind with Señor Villegrand. The latter urged his steed on at an ever increasing gait, with the utmost disregard for the character of the roads, so that I expected every moment to see the vehicle upset or the driver thrown from his seat and seriously injured. However, no such misfortune befell either, as we dashed along at a nine-mile gait for an hour or more toward our destination. So fast did we travel that there was little opportunity for making any observations concerning the nature of the country, or its fauna and flora. For a distance of some ten miles from Gallegos the road led straight across the almost perfectly level surface of a low, shingle-formed plain covered with scattered tufts of grass, which at this season was brown and withered. Suddenly we came to a depression some thirty feet in depth, which formed the bed of a dry water course leading from the plain to the Gallegos River. The road descended rather abruptly into this, and down into it plunged our team and driver without so much as an attempt at checking their speed. Having arrived safely at the bottom, a stop was made, in order to take advantage of the protection offered from the wind for lighting a cigarette. It was an unfortunate stop, however, for, the cigarette lighted, the cart-horse refused to go, and no amount of encouragement could induce him to change his mind. After some time spent in a vain endeavor to relieve the situation, I bethought me of the expedient of snubbing from the saddle horn, so often tried with signal success at home. Taking down the hard-twist saddle rope which, with my stock-saddle, I had wisely brought with me from the States, I attached one end of it to the cross-bar at the rear of the shafts, then taking a half hitch round the end of one shaft, with the free end of the rope in my hand I mounted my horse and, cautioning the driver to be in readiness for the start, swung into position directly in front of the cart horse, and taking a hitch with the rope on the saddle-horn, gave my horse the spurs and away we went out of the bed of the stream and up the incline on the other side. Once on top, Señor Villegrand felt renewed confidence, and signalled me to cast loose, which I had no sooner done than his horse stopped stark still and refused to go further, so that I was compelled again to resort to the same expedient, and I continued to snub the cart along by the saddle horn until within a half mile of Guer Aike, where, as we approached, we saw gathered a considerable assemblage of persons, brought together, as we afterwards learned, to witness the horse-races and other festivities which had been arranged for that afternoon and evening. When within a half mile of this place, Señor Villegrand, desiring to avoid the embarrassment of appearing in this somewhat helpless situation, again signalled me to cast him loose, which I immediately did, but much to his discomfiture the horse again refused to go, and the Chief of Police of the Argentine Territory of Santa Cruz had to submit to being towed into Guer Aike by a "gringo" Yankee, much to his mortification, I fear, and the amusement of the assembled group of gauchos, rancheros, runaway sailors, escaped convicts and other nondescript characters, who had gathered from the surrounding country to witness the races and take part in the drunken orgies following them.

We were to stop at Guer Aike over night, and had arrived in time to see the races, which had not yet commenced. Lest my readers be mistaken as to the size and importance of this place, which on the map appears quite as conspicuous as either Gallegos or Santa Cruz, I will venture a short description of it. Briefly, Guer Aike consisted of a single tumbledown building on the south side of the Gallegos River, at the head of tide water, some twenty-five miles from the mouth and at the first practicable ford on that stream. It was such a place as in our Western country in an early day would have been considered a road house of bad repute. It was at the time of our visit presided over by a Spaniard and his Chilian wife. The latter was a young, modest, rather handsome woman of prepossessing appearance, who seemed somewhat out of place and embarassed by her surroundings. There were in all three rooms, of which one served as the kitchen and living quarters for the family, a second was reserved as the sleeping quarters for the guests, while the third served as a dining room and *almacen*, where a stock of goods consisting for the most part of "watchaki," and other cheap and correspondingly bad liquors were displayed for sale or barter. Gathered about this miserable place was a motley assemblage of from fifty to one hundred men and boys, for the most part, even at the time of our arrival, more or less under the influence of liquor, all eager to witness the races for which they had assembled, and loud in proclaiming the special qualities of their particular favorites, each one apparently willing to back his judgment regarding the contestants in any particular event to the extent of his worldly possessions or credit. Most of those assembled were mounted, and their costumes were, to our unaccustomed eyes, peculiar if not striking, so that our curiosity was about equally divided between the people, their costumes, and the different manner of gear which served them for saddles on which to ride, no two of which seemed fashioned after exactly the same pattern.

Shortly after our arrival the crowd repaired to the race course, which was near at hand and consisted of a double track about a half mile in length, laid out on the surface of a perfectly level plain. The races were all run after the fashion in vogue in that country, which permits of fourteen false starts, "catorce partidas," before the final start. The contests were thus more in the nature of tests of the skill of the respective jockeys than the capabilities of the horses. They were exceedingly tedious, and I soon tired of watching them start and come perhaps half way down the course with every indication of making a close finish, when suddenly they would pull up and go back to the start to repeat again, after much delay, the same operation.

Leaving the race course and its votaries to their enjoyment, I wandered off across the level plain that lay between me and the river, in order to get a nearer view of the cliffs which formed the high embankment of the opposite or northern shore. As I approached the stream the ground was covered with a thick growth of *mata verde*, the beautiful dark green color of which at a distance gave the appearance of a field of growing grain or a stretch of luxuriant meadow land. This plant, which is by far the most abundant shrub throughout the plains of southern Patagonia, although resembling very closely in its foliage, manner of branching, strong resinous odor, the dwarf variety of juniper common in the arid regions of the West, is nevertheless a member of the Compositæ and in no way related to the juniper. Not only is the *mata verde* the most common, but it is also among the most useful of the shrubs of this region, for notwithstanding that it seldom attains a diameter of more than a half inch or a height of over three feet, through the great quantity of resinous matter which it secretes, when used as fuel it is of especial value, owing to its high calorific properties. While in Gallegos I noticed that it served as the only source of fuel used for domestic purposes by the inhabitants. Though, as may be gathered from my previous remarks, aside from culinary and laundry purposes, there was little demand for fuel of any kind.

After having strolled along the bank of the river for some distance, which at this point was separated from the main channel of the stream by a low mud flat entirely submerged at high tide, I turned to the southward, and crossing the wagon road by which we had come from Gallegos to Guer Aike, kept on until I gained the crest of a terrace some fifty feet in height overlooking the river valley. From the crest of this terrace a gently undulating plain extends inland, gradually increasing in elevation for a distance of from two to three miles, when it terminates at the foot of an escarpment forming a terrace similar to but somewhat higher than that just mentioned. Desirous of seeing as much as possible of the topography of the surrounding country, I continued my walk until gaining the summit of this latter elevation. Here at my feet there lay spread out before me the broad, low, level plain, which had at one time formed the valley of the Gallegos and Chico Rivers. Through the slow elevation that has taken place and is still going on along this coast, it was evident, as I stood and looked across the plain at the little village of Gallegos, that the entire tract of level country lying between the Gallegos and Chico Rivers had been but recently recovered from the sea, passing successively through the stages represented by the mud flats in and along the rivers, the shingle formations on the beach, the river valley lying between the latter and the terrace marking the boundary of the first of the series of narrow plains, on the escarpment of the second of which I stood. Not only were these plains covered with shingle, but the surface of this still exhibited, when carefully examined, slight elevations arranged in parallel lines very like those seen on the surface of a great bed of shingle now forming on the north shore at the mouth of the Gallegos River near Cape Fairweather, a photograph of which is reproduced in Fig. 1.

From my point of vantage I had a commanding view to the east and south. The surface of the low, level plain which lay at my feet stretched away to the eastward for a distance of twenty-five miles, interrupted only by the Rio Chico, and became finally lost in the deep blue of the Atlantic, while on my right the dull brown covering of withered grass met and blended at the southeastern horizon with the sombre gray of the low-hanging clouds. Far away to the south extended the same plain, inter-sected by the Rio Chico, while beyond this stream could be seen the black and broken surface of several lava streams, that had at no very remote time been poured out over this plain from a number of extinct craters, still visible in the distance. To the northward the field of vision was limited by the magnificent sea wall which rose abruptly to a height of four hundred and fifty feet from the bed of the stream and stretched almost uninterruptedly to the mouth of the river. Each stratum of clay, sandstone, conglomerate, or volcanic ash, was distinct and well defined, and as I looked from a distance upon this specimen of nature's handi-work, I could not avoid speculating as to what prizes in the way of skulls or skeletons of prehistoric animals were held for us locked in its stony embrace. I well knew that it was from the rocks of these same cliffs that Captain Sullivan had seventy years before collected the bones of certain fossil mammals, upon the scanty evidence of which Darwin had concluded, and rightly too, that the beds containing them were of more recent origin than the marine deposits forming the bluffs at the mouth of the Santa Cruz River. While sanguine of success, it must be confessed that we were not prepared for the almost embarrassing riches contained in the deposits which formed this line of cliffs.

The study of nature is always instructive and interesting, even inspiring and impressive, if the student be a real lover of nature seeking for truth at first hand and for truth's sake, and not merely a fireside naturalist, who seldom, goes beyond his private study or dooryard, and either contents himself, like other parasites, with what is brought to him, or like a bird of prey forcibly seizes upon the choicest morsels of his confrères, with little or no consideration for the rights or wishes of those who have brought together the material at so great an expense of time and labor.

We often hear and read of the monotony of the plains, and doubtless most of us to whom it has fallen to pass uninterruptedly any considerable period of time on the plains have at times distinctly felt this monotony. Nevertheless, there is something decidedly, though silently, impressive in a broad and level plain. As I looked about me I could not resist the drawing of a mental picture of those comparatively recent times when the waters of the Atlantic washed the foot of the escarpment on the crest of which I stood, and when the streams of molten lava were poured out over the surface of the great plain to the south. From this it was but another and comparatively short step backward to the time when, instead of a semi-arid region scantily clothed with vegetation just sufficient to sustain the present meagre fauna of these plains, this region supported a fauna rich in ungulates, sloths, armadillos, and giant flightless birds, all indicative of a mild climate and an abundant if not luxuriant vegetation. What a transformation had taken place since these animals had inhabited this region ! The southern Andes had been elevated to so great a height as effectively to deprive the prevailing southwesterly winds of their moisture while passing over the summits, causing them to descend the eastern slopes of these mountains and pass out over the plains in a desiccated condition, thus reducing the latter to their present semi-arid state and rendering them quite incapable of supporting more than a scanty flora and fauna. These, as well as many other physical changes, had taken place, which will be discussed later when we come to speak of the geology and geography of this region.

With the approach of darkness I returned to Guer Aike to find, that, while the races had been long concluded, the crowd had by no means dispersed. Some twenty or thirty still remained to make memorable our first Sunday spent on shore in Patagonia, through their drunken carousal protracted far into the night and emphasized by frequent quarrels. Toward morning these orgies gradually subsided as each participant became helpless from the effects of liquor and was stored away on the floor in a convenient corner by the proprietor, who generally succeeded in extracting from each his last dollar in return for the intoxicants which rendered him helpless, but at the same time less disagreeable.

Señor Villegrand was very solicitous as to our comfort and welfare, but despite his utmost efforts it was late in the night before we succeeded in getting anything to eat. At last an ample though not especially well prepared meal was served in the bedroom previously referred to, which had on this occasion been very considerately reserved for our party through the influence of Señor Villegrand.

Supper finished, we spread our blankets on the floor and turned in to pass a considerable portion of what still remained of the night in a vain endeavor to gain a little much-needed sleep. For some time this was quite impossible, but toward morning all became quiet and we fell asleep for a short time, though from the very nature of the situation it may be imagined that our slumbers were neither so sound nor restful as could have been desired. Always accustomed to rising early, we were up and dressed betimes on the following morning, and, on opening the door which lead from our room to the groggery, through which we had to pass to gain the outside of the building, there was presented such a sight of drunken and debauched humanity as beggars description and is best left entirely to the imagination.

Señor Villegrand and the others were not long in following our example, and apparently no less anxious than ourselves to be rid of the place. The horses were soon brought in, caught, saddled and harnessed. A different horse from the one used the preceding day was selected for the cart. In a short time we were ready to start, but, first of all, the Gallegos River had to be forded and we were not at all sure as to what would be the disposition of our new cart-horse after the experiences of the previous day. It was, therefore, with something of relief that we saw him start off all right. All went well until we had reached a position about midstream, when the horse refused to go any further. With commendable patience Señor Villegrand tried in every way possible to urge him forward, but he was absolutely incorrigible. This was indeed a situation, stuck fast in the middle of a stream of ice-cold water one hundred yards in width and from three to four feet in depth. After our driver had ineffectually exerted every effort to relieve the situation, and, sitting upright with the picture of despair in every feature, at last signified his willingness to permit me to do what I had from the first suggested, I rode alongside and attaching my saddle rope as on the previous day, though with greater difficulty, and with the additional assistance of a gaucho, who appeared with his horse at the opportune moment and fastened on to the opposite side, we succeeded in transferring safely to the other side of the river the horse, cart and driver.

Once on the opposite side our course lay for several miles up the narrow valley of a small watercourse, dry throughout the year, save in the early spring or for a few hours immediately following the heavy rains that do occasionally, though too rarely, visit this region. On our left was a low plain covered with a thin sheet of lava, the source of which could be seen in two small extinct craters rising somewhat above the surface of the plain and situated at a distance of some two or three miles to the westward of the road. On our right was a somewhat precipitous bluff extending from the bottom of the narrow valley to the surface of the broad and level plain some four hundred feet above. By following the course of this stream we finally emerged upon the broad level surface of the latter which lay spread out before us as far as the eye could reach ; its surface to the north and east seemingly unbroken by either elevation or depression, while to the westward there appeared in the distance the escarpment of a second terrace with a northerly and southerly trend, considerably elevated above the surface of the plain over which we were travelling.

After travelling for several miles in a northwesterly direction across this plain, we came suddenly to the brink of a great depression, some three hundred feet in depth and eight or ten miles in width. Owing to the southeasterly slope of the surface there was no indication of the presence of this depression until we arrived upon the crest of the escarpment surrounding it. At the bottom was a small salt lake called Laguna Leona (lioness lake) by the natives. In the spring and early summer months this forms a considerable body of water, but at the time of our visit in mid-autumn the volume of water had been greatly diminished through surface evaporation, aided perhaps by subterranean drainage, and there remained only the two ponds, the larger of which did not exceed a mile and a half in greatest diameter, shown in Fig. 2. The photograph was taken from a point about half way down the side of the bluff and does not do justice to either the height or abruptness of the latter. The trail zigzagged down the side of the bluff, which was so steep that Señor Villegrand had the horse taken out of the cart and the latter was taken down by hand, all lending their aid to the operation. At the bottom we followed along between the foot of the cliff and the shore of the lake, passing within sixty yards of a guanaco as he stood on a projecting bench and gazed at us as we rode past, with such temerity that I refrained from trying my Colt's new navy revolver,

1.—Bed of Shingle at Cape Fairweather.

2.—Laguna Leona.

FIGURES 1 AND 2—SEE OTHER SIDE

under the delusion that he was a pet much in favor perhaps with the children at the shepherd's shanty which we were approaching, where we stopped long enough to refresh ourselves with a lunch consisting of cold mutton and most excellent bread and coffee, hastily prepared and set before us by the Falkland Island wife of the shepherd who occupied the shanty. This frugal but ample lunch was none the less appreciated after our experience of the previous night, and, since it was our first refreshment for the day, we required no other relish than that provided by our twenty-mile ride over the pampa in the teeth of a stiff morning breeze, to stimulate our appetite to its full enjoyment.

Breakfast finished, I improved the short time allowed the horses for grazing in learning what I could of the basin, its lake and enclosing bluffs. On emerging from the cabin, a narrow strip of green grass was seen to commence near the top of the cliff and extend down its side and for a short distance out into the valley. So distinct was the brilliant green of this narrow strip from the withered brown of all that surrounded it, that it appeared as a green line ruled upon a brown surface. I at once divined its origin, and on approaching nearer was not surprised to see a beautiful rivulet of sparkling clear water dashing over its bed of many-colored pebbles on its way from its source near the summit of the cliff to the drouth-stricken valley beneath. This stream had determined the presence of the shanty and its inmates at such a seemingly out-of-the-way place. As I walked along toward the source of this little rivulet, I was struck with the distinctive character of the fauna and flora immediately adjacent to it. It was a miniature oasis in a vast, semi-arid, if not desert country. Instead of the dried-up and withered grasses of the plains there was an abundant growth of rich green grasses and sedges, with Equisetum, wild celery, and other moisture-loving plants fringing the banks, which were covered with mosses and Hepaticæ, while clinging to the rocks in the water were masses of Chara, beautiful green Algæ and other plants, in the branches of which swarmed myriads of amphipods, gastropods and bivalve molluscs, leeches, annelids, etc. Under the stones and in the grass along the margin were secreted a profusion of spiders, sow-bugs and beetles, belonging to those genera and species which are lovers of moist places. Nor was the change in fauna restricted to such small and inconspicuous forms as those just mentioned, for upon climbing the slope to the point, some thirty feet beneath the surface of the plain, where the spring gushes from beneath the bed of shingle overlying the tosca, when I sat down for a moment's rest, I was immediately struck with the variety and number of the birds about me, more especially since during our trip across the pampa we had been impressed with the scarcity of bird life. Nearby were a number of calafate and incense bushes, and about these were to be seen the red-breasted meadow lark, *Trupialis militaris*, somewhat larger than our yellow-breasted forms. The little crested song sparrow, *Zonotrichia canicapilla*, was everywhere about, as was also a small yellow-breasted finch, *Chlorospiza melanodera*, while in the dense foliage of the clumps of *mata verde* and *mata negra* that grew everywhere a little brown wren, *Troglodytes horhensis*, could be seen hopping about in the manner so characteristic of these birds. The several and deeply worn game trails that radiated from this spring bore unmistakable evidence that, prior to the time of the building of the shanty, this had been a favorite watering place for the bands of guanaco, rhea, and other larger mammals and birds indigenous to the region.

A hasty examination of the spring convinced me that the shingle was the source of the water supply, and the appearance of the latter at the surface indicated the point of contact between the shingle and the underlying tosca.

On turning my attention to the great depression before me, it was readily apparent that, while it was surrounded on all sides by high bluffs, these were decidedly higher to the westward than to the eastward, and by a subsequent study of this and hundreds of other similar depressions found scattered throughout the Patagonian plains I arrived at the conclusion that such depressions were once bays formed during the progress of the final recovery of this region from the sea and were formerly quite similar to San Blas, New Bay, San Julian and Peckett Harbor. As the elevation proceeded, such bays were transformed first into lakes by being cut off from the sea. Later, such lakes became entirely desiccated, or more or less reduced in size, according as the water supply from their tributaries compared with the loss by evaporation. The full discussion of the origin of these depressions will be left to the chapters on the geography and the geology of the region.

From the shepherd's house the trail led out across the valley for a few miles, then up the side of the bluff to the plain lying to the north of the depression, which was noticeably lower than that to the west. In a short time we had crossed this plain and, coming to its northern crest, looked down upon and across the valley of Coy River, or Rio Coyle, as it is locally known.

The valley of this stream has a breadth of from two to six miles and supports a vegetation consisting for the most part of grasses, which, as compared with those of the pampas, might be considered almost luxuriant.

The stream is small, seldom more than two rods in width, and usually sluggish in its lower course. There are long stretches of still and quiet waters alternating with occasional short rapids, as the confined waters dash for a few rods over coarse beds of shingle. We reached Governor Mayer's estancia, located some twenty miles above, early the following morning. Wherever we approached the river, it was noticeable that the surface of the water was but little beneath that of the valley, although this was the season of low water. From this it has resulted that the soil for a considerable distance on either side is subirrigated. This explains the presence of beautiful stretches of bright green meadow lands clothed with a luxuriant growth of tall waving grass. I later discovered that this was true of all the less important rivers of southern Patagonia, such as the Rio Sheuen, or Chalia, Rio Aubone, and the Rio Chico of the Gallegos, while the larger rivers like the Gallegos, Santa Cruz, and the Rio Chico of the Santa Cruz, are rapid streams with deeper channels cut in valleys that are as a rule quite as arid and destitute of vegetation as are the surrounding plains. This difference is evidently due to the fact that the larger streams have their sources far back in the Andes, where they are fed by numerous glaciers or mountain streams coming from the region of perpetual snow and hence have a constant and never-failing water supply, continuous throughout the year, with an erosive power sufficient at all times to keep their channels open. With the smaller streams it is quite different. They head on the plains or in the low foothills at the eastern base of the Andes. Their water supply is, therefore, precarious and intermittent, and, during the long dry season, falls so low that for long distances throughout their lower courses their channels are dried up and destitute of water, so that the materials brought down by the upper courses of such streams, instead of being carried out to sea, are thrown down just above where the flow of water ceases, resulting in the gradual silting up of the channel and the formation of the series of rapids or riffles and stretches of still water just referred to, through the inability of the stream to keep its channel free, owing to the reduced erosive power due to the diminished volume of water, which finally becomes negative.

We arrived at the estancia on the morning of May 5th, and the following morning Mr. Peterson, Señor Archerique and myself proceeded some twenty miles farther up the river to visit an Indian encampment. The day, like all the preceding ones since our arrival at Gallegos, was raw and cold, and rendered the more disagreeable by a piercing southwesterly wind. At a distance of a few miles from the ranch the river valley made a considerable deflection to the northeast, and, since we judged that our course lay to the northwest, we left the valley and climbed the bluff to the plain above. For miles we galloped over the barren, shingle-covered surface of this perfectly level tract. If the plain back of Gallegos had seemed somewhat destitute of vegetation, this was tenfold more so, and I was at a loss to understand what sustenance the guanaco and rhea, which appeared at intervals, could derive from the few almost leafless plants which lifted their short, fleshy, thorn-covered stems above the surface of the surrounding stones, or lay spread out in large, circular, cæspitose masses, presenting a surface almost as hard as that of the surrounding bowlders. The components of the shingle were noticeably coarser than were those of the plains nearer the coast.

It was evident from the topography of the surrounding country that the river made a considerable elbow here and that, by holding our course, we should again reach it at about the point where it resumed its northwesterly direction. After a most invigorating gallop of an hour and a half across the pampa, we came suddenly to the crest of the escarpment above the river. In the valley beneath grazed quietly a band of about three hundred variously colored horses belonging to the Indian village, consisting of half a dozen toldos pitched at the foot of the cliff some two miles farther up the valley. These toldos, made of guanaco skins sewed together and stretched fur side out over poles laid in crotches set in the ground, were not easily discernible, so perfectly did the brown fur of the guanaco skins blend with the brown of the grass-covered valley.

At first, as we approached, there was little evidence of life about the toldos. But a little later, as we drew nearer and our arrival had been announced, the full population of the village was in evidence, and we were greeted by from fifty to one hundred dogs representing every conceivable variety. These appeared so suddenly and in such numbers from all the toldos, and rushed upon us with such apparent fury, at the same time setting up such a pandemonium of savage barks, howls and growls, that for a time it looked as though they were intent on making a breakfast of us. The first alarm over, however, they disappeared almost as suddenly as they had appeared, so that we felt safe in dismounting and proffering the hand of friendship to the so-called giants of Patagonia. We were received with evident friendship and shown into one of the toldos, where yerba was being served, of which we were invited to partake, but declined. At this they took no offense, but continued in as friendly a manner as before, notwithstanding that other writers have stated that to refuse to take mate, or yerba, as it is called, is regarded by these Indians, and the Argentine gaucho as well, as a grave insult. During my three years' experience in Patagonia I found that these people were quite as reasonable as ourselves about such matters and freely accorded to all the privilege of declining this or any other favor offered.

Since the custom of taking mate is such a peculiar and prevalent one, not only in Patagonia, but throughout Argentina and much of South America as well, I shall diverge a little at this point in order to describe it. The beverage is made from the powdered leaves and stems of *Ilex paraguayensis*, a plant indigenous to Paraguay and portions of the interior of Brazil. Enough of this is placed in a cup, small gourd or other vessel to nearly fill it, then just enough cold water is poured in to moisten the contents, when a tube, usually made of brass, with a perforated bulb at the lower end, is introduced and pushed down to the bottom of the cup. A kettle with water is kept boiling over the fire and from this the cup is filled and passed successively to the different members of the party. Each person, on receiving the cup, sucks the liquid through the tube until the contents are exhausted, then returns it to the person who has volunteered to serve mate, by whom it is refilled and handed to another, and so on indefinitely. This process is kept up frequently for several hours at a time. But one cup and tube are used, and it is said to be the height of bad taste to wipe or in any way cleanse the tube before placing it in the mouth. So fond are the natives of this habit that they waste much valuable time and will neglect almost any work, no matter how urgent it may be, to indulge themselves in it. It is not a stimulant, I should say, but rather a sedative, and the claim is made for it by the advocates of its use that it arrests digestion and thus enables one to go for a longer time without taking food with little inconvenience. I have taken it a number of times and found it to be not particularly disagreeable to the taste. Whatever may be said of the medicinal or dietetic properties of the herb, the manner of taking it, as practiced in Argentina, is both filthy and conducive to dilatory habits, and therefore much to be deplored. In Argentina it might almost be referred to as a national custom, though it is likewise common throughout most of South America.

The Tehuelche village consisted of six toldos and some thirty individuals. They were for the most part well, or at least warmly, dressed in mantles made of the skins of the young guanaco, a small skunk (*Conepatus suffocans*) or a small cat (*Felis pajero*) indigenous to the country. The women were for the most part busily engaged in making and painting these mantles, while the men seemed to have little or nothing to do.

We remained some two or three hours at the village, deeply interested in the people and in their customs and arts, of which I shall speak further when I come to treat of the natives of Patagonia. After giving a few presents to the women and children and distributing a few packages of tobacco among the men, we remounted our horses and returned to the estancia late in the afternoon, in prime condition to enjoy a good meal of roast mutton, bread and potatoes, staple articles in the menu of every Patagonian sheep farm.

Late in the afternoon of the seventh of May we left the estancia, on our return trip to Gallegos. We had enjoyed the trip greatly. Everything was new and intensely interesting, but with it all there was a feeling that we should be making some progress in the securing of horses and outfit. I had seen enough of the country to be convinced of the possibility of taking a vehicle through almost any portion of it, provided only that we could procure such a vehicle and the necessary horses. We had passed two estancias and the shepherd's shanty on our way out from Guer Aike, and since Señor Villegrand was going to return by a cut-off, I decided to separate from the others and return the way we had come in order to see what success I might meet with in purchasing horses at these estancias, neither of which I found had any to spare. However, on the following day, when I returned to the shepherd's shanty, I succeeded in buying a very fair saddle horse for seven pounds, and after proceeding a few miles further I fell in with a Spaniard, one Francisco Cid by name, who was driving a troop of horses, from which I selected one, paying nine pounds. This gave us each a saddle horse, and going on to Guer Aike I found that our party had arrived in advance and put up for the night. Neither myself nor Mr. Peterson had any desire to pass another night at this wretched place, so saddling our newly purchased horses, we went on to Gallegos that night, leaving Señor Villegrand to follow in the morning.

Upon our return to Gallegos we applied ourselves quite energetically to the securing of additional horses, a suitable vehicle, and such other articles as were necessary for our equipment. Fortunately we had brought with us a tent, and a camp stove made of heavy sheet iron, such as we were accustomed to use in tenting on our western plains. The weather continued uncomfortably cold and disagreeable, and, in order to add to our comfort, we pitched our tent, put up the stove and secured a supply of wood, with which we made ourselves comfortable, while engaged in getting together, assorting and preparing our outfit. After dining in the cold, barn-like government building, it was a real comfort to repair to our canvas tent, the interior of which we kept delightfully warm.

We had been greatly perplexed as to the procuring of a proper vehicle. That offered us by Señor Sanchez was entirely too low and too heavy for our purposes. For a time, however, it appeared that it was to be that or nothing. Just as we were about giving up in despair, Señor Ferrari, an Italian and the keeper of the principal shop in the village, came to our rescue by very considerately offering us at a moderate price a two-wheeled horse-cart fitted with shafts and a shaft harness, which he had been using in conveying his goods from the beach to the store. This was really very kind of Señor Ferrari, considering that the vehicle was almost as much of a necessity to him as to ourselves, and the trouble and delay he was put to in having another brought down from Buenos Aires. Needless to say, we accepted this offer at once, for while the cart was by no means what we could have wished, it was by all odds the best we could do under the circumstances. We now had cart, harness and saddle horses, but had yet to procure cart-horses.

Horses are so seldom broken to harness in Patagonia, the carting for the most part being done with bullocks, that the possibility of procuring well-broken cart-horses was quite out of the question. However, an Argentine gaucho and Indian trader, who had been present when we purchased the cart, assured us that he had a good, gentle cart-horse which he would bring and show us the next day. In accordance with his promise he appeared on the following morning with a considerably undersized and rather inferior looking animal, which on examination proved to be only three years of age and had evidently belonged to some Indian family with children, by whom it had been ridden and driven almost from the time of its birth, which was the cause of its stunted and unthrifty appearance. Our friend, Señor Ortis by name, proceeded immediately to harness and hitch to the cart "El Moro," the roan, in order to display to us his many good points as a cart horse. All being ready he jumped in the cart, seized the lines, and yelling and whipping incessantly circled about over the rough and uneven country lying between us and the beach at a most reckless gait, continually urging "El Moro" to greater efforts and occasionally drawing near and halting for an instant in order to tell us that *el moro era un buen caballo con mucha fuerza*, when he would again dash off at a frantic rate, as though the merits of the animal were entirely dependent upon the greatest speed he could attain. This operation was repeated several times, when out of compassion we purchased the animal, which, however deficient he might be in physical strength, was evidently willing. "El Moro" justified our expectations and proved a tractable, docile and willing animal.

It is the custom in Argentina, in working horses to carts, to have only one horse in harness, which of course is the shaft horse, while on either side is placed a cinch horse. These pull by a single cinch rope and from one side of the animal only. One end of the cinch rope is made fast to the cart at any convenient place, while the other is provided with a hook which is hooked into a cinch ring on that side of the horse next to the shafts, the cinch having previously been made taut and the body of the animal protected with rugs and a pair of heavy leather pads stuffed with straw, some eighteen inches long, oval in cross-section, about two by four inches in diameter, placed one on either side of the horse's spine.

We were not long in procuring two mares, which we used as cinch horses, and the morning of May sixteenth found us ready and equipped for our work, just two and one half months after our departure from New York.

CHAPTER III.

We start on our collecting trip; Abundance of amphipods in fresh water springs; Detained at Guer Aike by a rise of the Gallegos River; Searching the bluffs of the Gallegos at Guer Aike for fossils; Visit to Killik Aike; Fossils abundant; Hospitality at Killik Aike; Habits of carrion hawks of Patagonia; A flock of Rhea; New world of animal life in the fossils; Move camp to North Gallegos; Fitzroy's Springs; Cape Fairweather; Our first guanaco; Peculiar antics of these animals when frightened; Methods of the Carranchas in attacking a carcass; The Condor, its mode of flight and attack on a guanaco carcass; Discovery of the Cape Fairweather beds, a new geological horizon; Landslides along the sea cliffs; Some of the commoner birds near Cape Fairweather; Habits of Canis azaræ; Peculiar experience with a condor; Hardiness of some of the Patagonian Compositæ; Cañon de Palo; The finding of the carcass of a whale.

SUCH of our equipment as was not immediately needed was stored in Gallegos, while our tent, stove, bedding, collecting and packing materials, necessary tools, ammunition, and a supply of provisions sufficient for a month, were placed in the cart, and, thus equipped, we left Gallegos on the afternoon of the sixteenth and camped that night at a spring some five miles from Guer Aike. It was a novel experience to be traveling through the country with this singular outfit, considering all the conveniences for overland travel to which we had been accustomed at home. However, we were determined to make the best of it and resolved that on our next trip to Patagonia we would bring with us a vehicle adapted to our requirements and the country. We were really surprised to see how well we got on with our nondescript outfit, though it took us some time to familiarize ourselves with the gear and learn the horse language of the country. The latter, I am afraid, we never did properly master. It was late in the evening when we arrived at our camping place, and as I went to the spring for water with which to prepare supper,

I was amazed at the extreme abundance of a certain species of amphipod, *Hyalella patagonica*. These little crustacea were present in such numbers that, in places over the sides and bottom of the pool, they appeared as animated masses from one to two inches in depth and several inches in diameter. Nowhere else did I see them in such numbers, though they were everywhere the most abundant and omnipresent forms of animal life in the fresh waters of the region.

Supper over, we spread out our beds and retired for the night without other shelter than that afforded by our tarpaulin, which, however, was ample. It was to both of us a great relief to be able to feel that we were at last in a position to commence the work which we had undertaken and looked forward to with so much anticipation. On the following morning we resumed our journey, intending to cross the Gallegos River at Guer Aike and establish a permanent camp on the north side at some point from which we could conveniently examine the exposures shown in the lofty cliffs, which there formed the left bank of the stream, for vertebrate fossils. On arriving at the crossing, however, we discovered, much to our dismay, that the stream had risen to a height sufficient to render it impassable with a cart. We therefore decided to camp temporarily a little distance above the ford and devote our attention to the immediate vicinity, until such time as the water in the river subsided sufficiently to permit of crossing. Selecting a suitable camping place, we pitched our tent, put up our stove, and prepared to make ourselves comfortable for whatever time fortune might necessitate our remaining at this locality. After partaking of a hastily prepared lunch, I saddled my horse and rode away up the river to examine some exposures which were distinguishable a few miles distant, while Mr. Peterson occupied himself in setting a number of traps for small rodents and in procuring and skinning representatives of the bird life of the vicinity. The exposures mentioned proved to consist of rather fine-grained, light-colored sandstones, composed largely of volcanic ash, exhibiting frequent examples of cross-bedding, though remarkably destitute of vertebrate or other fossils. I returned to camp late in the evening, not forgetting however to stop and break down a bunch of dead calafate bushes, *Berberis cuneata*, which I had detected shortly after leaving camp. These, when dragged into camp with my saddle rope attached to the horn of the saddle, furnished fuel sufficient to enable us to pass a comfortable evening in the tent about a cheerful fire which burned brightly in our stove, while the bitter south-west wind without swept madly over the bleak and barren plain and strove with all its energy to overturn our tent, causing the guy-ropes to strain at their anchors and give forth those peculiar musical notes with which we had become only too familiar through our long experience with Wyoming blizzards and the so-called zephyrs of the Kansas plains. On the following morning, the river being still too high to cross with a cart, I resolved to go over on horseback and spend a couple of days at least prospecting the very promising exposures on the other side for vertebrate fossils, Mr. Peterson meanwhile remaining where we were, to look after the outfit, and continue his work of collecting birds and mammals, in the prosecution of which he was meeting with considerable success. It must be confessed that the stream did not look particularly inviting, as its icy cold waters rushed along with a seven- or eight-knot current. However, while the water was uncomfortably cold, the stream was not particularly dangerous to one accustomed to such streams, and I crossed in safety the water in the deepest portion of the channel just flowing over the top of my saddle seat. Once on the other side I removed my lower clothing, wrung them out as best I could, replaced them, and proceeded to the foot of the nearest bluff, where, leaving my horse and taking my pick, I began a search for fossils. The cliffs at this point have an altitude of four hundred and fifty feet, and are composed of alternate layers of sandstones and shales, the former predominating. The sandstones are largely made up of fine volcanic ash, though there are occasional beds of only local extent of conglomerates, consisting for the most part of volcanic materials. Several hours' search at this locality was only rewarded by the discovery of a number of unimportant fragments. I therefore decided to return to my horse and go to Killik Aike some twelve miles farther down the river, where I had been told fossils were to be found in considerable abundance. There was an estancia at Killik Aike owned and operated by Mr. H. S. Felton, to whom, as well as to various other estancieros of the country, Governor Mayer had very kindly and thoughtfully given us letters of introduction, before we left Gallegos. I arrived at Killik Aike shortly after midday, and as Mr. Felton was absent from the estancia, I presented my credentials to his foreman, at the same time stating the purpose of my visit and asking if I might be given accommodation for a day or two, during which to make some preliminary examinations. I was most hospitably received and invited to make myself at home for as long a period as I saw fit.

Killik Aike lies at the bottom of a deep gulch, which enters the Gallegos River from the north. At its mouth this gulch is not more than three hundred yards in width, but at a short distance above it opens out into a wide basin, entirely surrounded by the high and level plain, which, however, is marked by numerous deep indentations, scooped out by as many small and usually dry water courses, tributary to the main basin.

I was not slow in accepting the invitation to make myself at home, and immediately unsaddled my horse and turned him out to graze with the other horses about the estancia. Then, after refreshing myself with a hearty meal of meat, vegetables and coffee, I set out for the beach, only a hundred yards from the house, where the foreman assured me there were to be found in the rocks many bones and teeth. I had scarcely clambered down over the edge of the three or four yards of silt, which forms the bottom of the small tributary valley, when at my feet, in the solid tosca, I discovered the jaws and teeth of a small rodent, *Procardia elliptica*. These were soon taken up and carefully wrapped in cotton batting. Turning to my right I proceeded to walk along the foot of the cliff, which fronts the river and increases rapidly in height, until, within a short distance, it rises perpendicularly to the plain above at an altitude of four hundred and fifty feet. I had gone but a short distance when my eye caught the reflection given off by the polished enamel of a tooth protruding from the surface, at a short distance from the base of the cliff. Upon examination this proved to be one of a complete series belonging to an almost perfect skull of *Icochilus*, a small ungulate mammal with hypsodont teeth, belonging to the Typotheria, which, though represented by numerous genera and species in these deposits, has left no descendants among the present fauna of South America, or elsewhere. The discovery of more or less complete skulls and skeletons of other animals followed in rapid succession. Among the more remarkable may be noted Nesodon, an ungulate mammal approaching in size some of the smaller rhinoceroses; Astrapotherium, another ungulate, and the largest and most formidable of the entire fauna, with its enormously developed upper and lower canine teeth and powerful posterior molars resembling those of the modern rhinoceros; Diadiaphorus, the horse-like ungulate, which has so many perissodactyl characters similar to those found in the modern horse, that, by some anatomists, it has been erroneously considered as ancestral to the latter animal; Prothylacinus, a large carnivorous animal exhibiting in its skeleton so many marsupial characters that it has very generally come to be considered as ancestral to the Tasmanian wolf, *Thylacinus cynocephalus*, as its name implies. Remains of these and many other mammals were to be seen protruding from the face of the cliff, or weathering from the surface of the great blocks of rock that had broken away and fallen to the bottom, where they were being rapidly disintegrated by the action of the tides. Truly this vast cemetery, which for untold ages had served as nature's burial ground, was now being desecrated by her own hand, with no one present to remonstrate against her wanton destruction of those remains whose very antiquity, it would seem, should have insured them against such desecration. We are wont to speak of the kindly hand of nature and attribute all our physical ailments, at least, to a disobedience of her laws. As I stood that afternoon and calmly viewed the surrounding scene, I could not help doubting the full truth of both assumptions. Look whither I would, the scene was one of utter desolation, while the very magnitude of the destruction which was being accomplished on such an enormous scale all about me was emphasized by the power of the tide, as it came rushing in great waves up the river, covering the broad mud flats and rolling in among the bowlders at my feet, compelling me to seek safety at successively higher elevations, where I could at last sit comfortably and look across a body of water of a depth sufficient for the safe navigation of ocean-going vessels, and where but a few hours previous there had existed only broad, level, mud flats, separated by unimportant streams, which I later crossed on horseback without inconvenience. The effectiveness of these tides as erosive agents was forcibly shown by the enormous power exerted by the great tidal waves as they came plunging forward, frequently presenting a solid front eight to ten feet in height and several hundred yards in length and rolling great stones of from one to two hundred pounds weight up and down the beach and rocking others weighing a ton, or even more, back and forth, when not in a position of perfect equilibrium, each successive wave carrying away great quantities of material on its retrograde movement. In this manner the four hundred and fifty feet of rocks forming the cliffs which towered above me were being rapidly eaten into and destroyed, and as I looked across the river and away to the south over the low plain which stretches for miles toward the eastern entrance to the Straits of Magellan, the enormous amount of destruction already accomplished was apparent. The very magnitude of the scale upon which the work was being accomplished commanded admiration, but after all, was there not a certain resemblance between this constructive and destructive work of nature and that of the speculative sciences, where old theories promulgated after long and patient research are for a time accepted as truth, only to be demolished later by the discovery of new facts?

Turning to the fossil remains about me, it was evident not only that they all belonged to species long since extinct, but that many of them had left no descendants. As I drew a mental picture of the physical conditions of this region in those remote times, when these animals roamed over, lived, died and left their bones on the uplands to bleach and be consumed, or in the lowlands to be preserved in the muds and sands about the shores, over the flood-plains, or in the bottoms of the rivers, lakes and marshes, I could not help wondering what had caused the extermination of this great fauna. As I examined the remains of *Icochilus*, *Nesodon*, and *Prothylacinus* that lay before me, they appeared well adapted for maintaining themselves even under the semi-arid conditions at present prevailing. The first two are supplied with incisor teeth which must have been very efficient in cropping even the shortest grasses, while the molars were admirably adapted for masticating purposes. As a carnivore *Prothylacinus* would appear well equipped for sustaining itself. Was their extermination due to a certain lack of vitality inherent in the different genera and families, whereby they became extinct through a want of fertility due to decreased vitality? In other words, does a species, genus, or family, like the individual, pass through the different stages known as birth, youth, maturity, old age and death, regardless of environment, which may accelerate or retard, but cannot prevent, in the former case any more than in the latter, the final dissolution, no matter how congenial the environment may be? Or was their extermination due to some great calamity which overtook them, perhaps suddenly, when at the zenith of their development? It is perhaps more likely that their extermination was due to gradually changing physical conditions, resulting in changes of climate, vegetation, distribution of land and water and other environmental conditions, to which they were unable to adapt themselves.

This first afternoon spent in the fossil beds at Killik Aike had in it all the elements of success. The intense and absorbing interest connected with this sort of work is known to but few and must be experienced to be understood; it cannot be explained. Besides the deep scientific interest connected with it there are often all the painful anxieties and uncertainties of the prospector after precious metals, and the collector of vertebrate fossils is often sustained for weeks or months at a time, it may be, by the same indomitable spirit and forlorn hope which impels the prospector to take his meagre grub-stake and spend months in the mountain solitudes in search of hidden treasures. Indeed the simile may be carried still further, for both are too often deprived of their fair share of the gains that follow, which in the one case take the form of money and in the other that of credit for work done.

I returned to the estancia late in the evening with a number of good skulls in my collecting bag, and several promising finds located, to be developed later. And what was more consoling than all, the success of the expedition I now felt to be assured. After a plain but substantial dinner I was invited by the foreman of the estancia to his room, where I spent the first really comfortable evening since our arrival in Patagonia. Mr. Felton and his wife, though absent, were evidently people who believed in enjoying some of the comforts of life, even though they did live at the other end of the world. There was an open grate in which glowed a cheerful fire of good English coal. The house was well furnished with comfortable chairs, a piano, a well-selected library and numerous periodicals of the latest numbers, brought from England on the bimonthly mail steamers which touch at Sandy Point *en route* for the west coast. Nor were our host and hostess unmindful of the inner man. There was an abundant supply of provisions and a choice selection of liquors, wines, beer, ale, stout, and mineral waters. Everything about the ranch bespoke comfort and consideration for the family, employees, and guests. We later had the pleasure of making the acquaintance of Mr. Felton and his charming wife and daughter, and many of our most pleasant remembrances of Patagonia are connected with the friendship and hospitality which they so freely and generously extended to us.

The foreman was a young Londoner, who had a great deal to tell me about that city, and wished to learn more about the States, so that the night was far spent when he showed me to my room, where I passed a most comfortable night on much the best bed I came upon during my travels in Patagonia. I arose greatly refreshed the following morning, resolved on rejoining Mr. Peterson at Guer Aike, in order to proceed with our entire outfit at once to Killik Aike, provided the river had fallen sufficiently to permit our crossing it with the cart. Taking with me the skull of *Icochilus* and one or two of the other smaller specimens, after coffee I caught and saddled my horse and set out for Guer Aike by a route by way of which I had been assured by my host that it would be feasible to return with the cart. On my arrival at the crossing of the river I found the water still high, but falling rapidly and with every prospect of our being able to ford it safely on the following morning. I immediately crossed over and proceeded to our camp, where I found Mr. Peterson busily engaged in preparing the skins of a number of birds and small mammals taken the previous evening. When I laid before him the few fossils I had brought as "samples" and detailed to him the remarkable richness in fossil remains of the exposures I had visited, his anxiety was no less than my own to be off at once and establish a comfortable camp somewhere within easy working distance of this newly discovered El Dorado. We immediately set about putting things in shape for moving the following morning, by which time it appeared that the water in the river would have subsided sufficiently to permit of our crossing. As to this we were not disappointed, and, after an early breakfast, on the next morning we were off. The river was crossed without accident, and, as we pursued our course across the low valley on the other side, winding in and out between the numerous bog-holes and small marshes that dotted its surface, our curious equipage caused no little commotion among the upland geese, *Chloëphaga magellanica*, which at this particular season flocked here in such numbers as literally to cover the margins of the swamps and small pools, where they feed on the tender shoots of grass which appear above the surface of the cold but humid soil. These birds were not especially shy, though always removing to a respectful distance when approached. Quite different were the carranchas, or carrion hawks, which, as we approached the estancia of Woodman and Redman, located at a distance of about a mile from the crossing, were to be seen in great numbers, eagerly devouring the carcasses of the dead or dying sheep, as they lay scattered about in considerable numbers over the valley. These birds seemed as absolutely fearless or regardless of our presence as are the fowls about a country barnyard. While driving along I frequently noticed that, when by chance a carcass lay directly in our way, instead of taking to flight at our approach, they would linger until almost directly beneath the horse's feet, when they would reluctantly hop off to one side in their peculiarly grotesque manner, uttering, at the same time, a series of loud, harsh, rasping notes as they stood at a few feet only from our cart, as though to remonstrate with and scold us for this apparently unnecessary disturbance of their feast, which was quickly resumed immediately we had passed by.

There are three species of carrion hawks in southern Patagonia. These pertain to three different genera, *Polyborus tharus, Ibycter albigularis,* and *Milvago chimango,* but are all given the common name of carrancha by the natives. I could not detect that they had any distinguishing names for the different species, which, though of almost identical habits, were so unlike in color. Although frequently seen mingled together and feeding from the same carcass, it cannot be said that such social relations were entirely harmonious, since they indulged almost continuously in most spirited personal conflicts, though I cannot assert that such engagements were any less frequent between individuals of the same species than between those belonging to different species.

Carranchas are quite common all over the Patagonian plains and in the depths of the Andean forests as well, where they seem equally at home. Indeed, being true scavengers, their distribution and numbers at any particular place would seem to be directly proportionate to the food supply. They are therefore found in considerable numbers in the forests of the lower Andes and along the water courses to the eastward, somewhat rarely on the high, arid, lava-covered plains lying between the Andes and the Atlantic, while they fairly swarm about the estancias near the coast, where they feed on the dead carcasses of sheep, horses or other animals, the former of which die by the thousands from old age. Through the lack of proper facilities for the exporting of mutton from this country ,the wethers as well as the ewes are kept for their wool, until they die from senility or disease. Thus to one accustomed to sheep raising in other countries the percentage of loss among the flocks appears abnormally high, as he sees hundreds or even thousands of carcasses lying about the paddocks and dipping pens, or scattered over the pasture lands. By devouring these carcasses the carranchas, aided by the condor, *Sarcorhamphus gryphus,*

and certain species of gulls, render to the inhabitants of Patagonia a service of inestimable value in maintaining the present most excellent sani-
tary conditions.

Our trail led past the settlement of the estancia mentioned and for a short
distance up the same cañon we had followed on our road to the Coy
River, then diverged, leading up the course of a small lateral cañon which
entered from the east, finally emerging upon the broad and perfectly level
plain which, with an elevation of four hundred and fifty feet, forms the
summit of the cliffs extending along the river front. The surface of
shingle was partially covered with soil, which gave sustenance to short
but highly nutritious grasses. Scattered over the plain at intervals were
to be seen flocks of sheep quietly grazing, in evident satisfaction with
their surroundings, while here and there a solitary guanaco might be seen
engaged in the same occupation and, attracted by the unfamiliar appearance of our equipage, would raise to its full height his long neck, which,
after intently surveying us for an instant, he would throw into a most
graceful curve, at the same time uttering a series of sharp, shrill
whinnies, intended apparently as a salutation, very similar in tone and
duration to those I have frequently heard given forth by a horse, when riding by at a little distance. After repeating this noise a number of times
he would resume feeding and become apparently perfectly oblivious of, or
indifferent to our presence. Occasionally a specimen of *Canis azaræ*,
a small gray fox or dog, might be seen dashing madly across the plains,
if by chance we came so near as to frighten him from the clump of
mata verde or other shelter behind which he had lain concealed. This
is withal a most beautiful, intelligent, and amusing carnivore, but at the
same time, as we were to learn later, both destructive and mischievous.
Scattered at intervals over the plain were small thickets of calafate bushes,
the trunks seldom more than an inch and a half in diameter and crowded
closely together. These usually attained a height of six or seven feet,
and the thick branches at the top were the favorite nesting places for the
carrancha, which constructs its nests of great quantities of sticks and stems.
I have frequently seen several bushels of such materials brought together
in a single nest, while the depth of the depression for the latter is out of
all proportion to the great mass of material employed in its construction
and only just sufficient to insure the eggs and young from rolling out and
falling to the ground.

After a couple of hours' travel across the broad and level surface of the pampa we came to the head of a long draw leading down to the cañon, at the mouth of which is the Killik Aike settlement. Hardly had we started to descend this draw, when we came suddenly upon a drove of ostrich, *Rhea darwinii*, numbering some fifteen or twenty. They were the first we had seen except at a distance, and it must be confessed that our inspection of these at close range was not very satisfactory, for immediately they had seen us they were off like a shot at an incredible speed, only equalled by that of the fleetest horse or hound. While the rhea is a flightless bird, its locomotion is nevertheless materially aided by the wings, which, when running, are kept moving in such manner as to aid in propelling the animal forward, or are held stationary in a partially expanded position, so as to act as sails when running before the wind. We arrived at Killik Aike in good season, and, selecting a suitable camping place near a spring a few hundred yards above the settlement, we pitched our tent, ditched it well, and banked it up around the bottom, hauled in a supply of wood, and prepared to make ourselves as comfortable as the circumstances would permit during the time necessary to complete our work at this locality.

So abundant were fossils in the exposures along the bluffs of the river at Killik Aike that we were kept busy, working early and late and seven days to the week, for nearly a month. First we worked the bluffs which I had prospected on my first visit and which lay up stream from the mouth of the cañon. Then, when the supply of fossils here became practically exhausted, we turned our attention to the cliffs below the cañon, which were discovered to be equally rich in fossil remains.

While the season of the year was that at which in this latitude the days are shortest and the nights longest, nevertheless, on account of the lingering twilight, we were able to accomplish a good day's work. We breakfasted each morning long before sunrise, and when that messenger of day appeared through the gray morning mists, rising like a brilliant red disc out of the waters of the Atlantic on our eastern horizon, we were already miles away and hard at work completing the excavation of some skull or skeleton commenced the preceding day, or perchance greatly elated over some new and important discovery. The very wealth of the material about us was an incentive to greater efforts. We were continually encountering and becoming familiar with an animal life pertaining to an entirely new world, the like of which we had never seen before. Not only were the genera and species distinct from any with which we were previously acquainted, but the families and orders were also, for the most part, quite different from those with which we had become familiar in the North. It was like being transferred suddenly to a new world peopled with a new fauna entirely different from that of our earth. Little wonder therefore that we put forth every energy to make the most of our opportunity in the midst of such a wealth of material and surrounded by such exceptional incentives. The work began in the early morning, was carried on until long after sundown, interrupted only for a few moments at midday while we ate our frugal lunch. Then, as twilight began to disappear, giving way to the greater obscurity of night, laden with the treasures rescued during the day safely packed in our collecting bags, we would dismiss from our minds those subjects with which we had been so intently occupied and find time to realize that such work is not only exciting, but exhausting, as we dragged ourselves wearily into camp, where, after a hearty meal, we would retire for the night and give ourselves up to that deep, restful and refreshing sleep, which to myself is unknown when living under more artificial, though some might say less savage conditions. It was the sleep of my childhood returned to me and it brought with it pleasant dreams of those happy days when, as a child, I roamed over our own western plains with a freedom almost equal to that of the birds and animals about me.

We continued our work at Killik Aike uninterruptedly until June 19th, when we carefully packed the ton and a half of fossils we had collected and shipped them by the "Bootle," a British schooner which had come to Killik Aike to load wool for London and was then lying at the mole. It was rather a remarkable sight to see this ocean-going schooner come up the river at high tide, sailing over what had but a few hours previously been mud flats and take its position at the end of this little mole where, twice every day, while receiving cargo, it would be left high and dry by the receding tide. After loading our boxes on the "Bootle," we moved camp farther down the river to the estancia of Mr. William Halliday, at North Gallegos, directly across the river from Gallegos. We remained here a week working up the river to the point to which we had worked down from Killik Aike. At this place we packed a large box of good material and as the "Bootle" came down and was beached here to complete her cargo, after finishing at Killik Aike, we shipped this along with the others on June the 26th.   Thus not quite four months after leaving New York our first consignment was on its way home, and, if safely received, the material was ample to insure the success of the expedition from a palæontological standpoint, which had been our real aim.   We had, however, not entirely neglected other branches of natural history and already had a considerable collection of bird skins, as well as a few mammals.

Since we were now in the middle of winter, we decided to pay more attention to the recent fauna and secure, as nearly as possible in full winter plumage, a complete set of the various mammals and birds that wintered in the region about us.   In prosecuting this work we intended still to carry out our plan, formed at the start, of following the coast and examining the bluffs which extend from Cape Fairweather northward, in order to study their stratigraphy and collect such fossils as they might contain.

On June 26, we moved camp to Fitzroy's Springs, about a mile above the mouth of the Gallegos River, where, under a high bluff and in the midst of a considerable thicket of bushes, we camped for the night, going on the next day to a small spring which flows from near the summit of the cliff that faces the Atlantic and runs northward from Cape Fairweather, distant about one mile from the extreme point of the cape.   We established our camp in a little basin on the plain above, but conveniently near the spring.   The surroundings appeared promising and we therefore selected and prepared our camp for a considerable sojourn at this place. The country for some two or three miles to the northward of Cape Fairweather consists of rounded hills separated by shallow depressions, while a loose, warm, sandy soil takes the place of the shingle of the plains and supports a considerable growth of grass and calafate bushes.   It is a favorite retreat for the guanaco and rhea, as well as many other mammals and birds.   We spent the remainder of the day after our arrival at this place in putting our camp in shape, providing a supply of wood, and doing such other little chores as would insure our future comfort.   On the following morning we took our Winchesters and started out after guanaco, as we wished to secure while here convenient to Gallegos a sufficient supply of skins and skeletons of these animals, more especially as their fur was at that season in prime condition.   After proceeding a little way together we separated, and seeing a small band of guanacos on the great bed of shingle that lies just above the mouth of the river I immediately directed my course toward them and walked deliberately up until within eighty or one hundred yards, when, selecting a pair that appeared especially fine, I fired two shots in rapid succession from my 45–90, bringing both to earth almost in their tracks, while the others, startled by the report of my weapon, made off across the bed of shingle in their peculiar swinging gallop, continually dodging first in one direction and then another and all the while swaying their long willowy necks up, down, and sidewise, at one moment raising their heads high in air, while an instant later they would duck suddenly until the nostrils were brought almost on a level with the ground.   After watching their peculiar antics for a moment, I proceeded to the scene of my destruction in order to get a closer view of their dead or dying companions.   Lying almost side by side on the surface of the clean and polished shingle were the two splendid beasts, representatives of the largest of the Patagonian animals.   As I approached they stretched their long necks toward me, while at the same time from their strikingly large and beautiful eyes they looked up, as though imploring mercy.   I almost wished I had not fired the fatal shots.   Since, however, I had been the cause of their destruction, I would not be unnecessarily cruel and leave them to a lingering death, so, reaching for my knife, I drove the glistening blade between the atlas and occiput quite through the spinal cord and deep into the cranial cavity of each.   In either instance there was a sudden convulsion of the entire body, followed instantly by a stiffening of the limbs and all was over.   My mission of mercy finished, I looked about me and saw that already a number of carranchas had gathered.   Several of these were standing about on the shingle within only a few feet of me, anxiously watching the operation, and, judging from their countenances, carefully calculating the length of time I should require before yielding the floor to them.   Unfortunately I was forced to do this earlier than I had anticipated, for, on preparing to skin the guanacos, I discovered to my chargin that I had nothing with me with which to make the necessary measurements of the dead animal before removing the skin.   While considering the proper course to pursue I saw Mr. Peterson walking at a distance of only a half mile, and signalling him to stop, I drove away the carranchas and walked over to him in order to borrow his tape line.   We immediately returned together only to find that in this incredibly short time the carranchas had not only returned, but had so mutilated the skins of both animals as to render them unfit for mounting purposes. The eyes and tongues were considered the chief delicacies, for not only were the former entirely gone when we returned, but an incision had in each instance been made through the skin at the base of the mandibles and the tongue of each detached and carried away, together with some of the hyoid bones, thus rendering the specimens unsuitable for skeletal purposes. The despatch and neatness with which the tongues of these animals were removed were simply marvellous.

If I had lost my first two guanacos, killed in Patagonia, I had at any rate gained at first hand some very useful knowledge concerning the habits of the carranchas, those scavengers of this part of the southern hemisphere; and since the carcasses were no longer of value as natural history specimens, I resolved to see what more of interest they would be the means of teaching me concerning the habits of these or other birds or mammals. Repairing to a thicket of calafate bushes that grew on the margin of the bed of shingle, where I could watch without attracting attention, I took a comfortable position in order to observe the drama which was being enacted on the surface of this bed of shingle, in the first act of which I had been one of the principal participants, though now reduced to the rank of a spectator. What an animated scene was transferred suddenly as if by magic to this particular spot, which under ordinary circumstances is quite destitute of either animal or vegetable life. Carranchas gathered about in great numbers, while several white and gray gulls also appeared upon the scene. Many a spirited contest was waged between these various aspirants for the more desirable portions. Suddenly there was on all sides a scurrying, as these birds literally tumbled over one another in their haste to get away, and looking up I saw a great male condor come sailing down. His powerful wings were fully extended, giving a total expanse of not less than nine feet. The long primaries were motionless, seemed separated from one another by an interval of about an inch and were each so distinct that I could without difficulty have counted the series as he soared by. The pure white of his shoulders and the delicate ruff was intensified by the deep black of the back and under-body, while the iridescence of the neck and comb were faintly distinguishable. He sailed straight as an arrow to the nearest carcass of the two guanacos, and on reaching a point directly over and a few feet above the dead animal, the limbs and feet were dropped suddenly downward to their full length, there was a slight quivering motion, better described perhaps as a tremor of the wings, and the great bird settled slowly downwards until the feet came in contact with the carcass, when the wings were stretched slowly to their greatest expanse and then carefully and slowly closed upon the sides. After alighting in this somewhat deliberate manner, a moment was given to a survey of the immediate surroundings, either for the purpose of detecting an enemy, or perhaps to confer most gracious recognition upon those other members of the feathered tribe who had at his approach by common accord moved off and now stood at a respectful distance, patiently awaiting the pleasure of this monarch of the air as he occupied not only the head, but all of the table at this most bounteous feast. This survey completed and having apparently convinced himself that all was plain sailing, he strode quietly up and down the body of the animal, once or twice, as though to select the most advantageous position, then coming to a standstill with the legs far apart and one foot planted firmly on the hip and the other on the side he commenced a most vigorous, even furious attack upon the carcass, using his great sharp and hooked beak as a weapon. Almost instantly a hole was made through the flank and into the abdominal cavity, where at each stroke with the powerful beak considerable sections of the intestines and other organs were torn out and eagerly devoured. After watching this performance for a few moments I bethought me of my Winchester, and mindful of what an excellent specimen he would make to adorn the ornithological hall of the new museum at Princeton (when some friend of that institution should find it in his heart to give the funds necessary for the construction of that much-needed building), I took the trusty weapon, adjusted the sights for a two hundred yard range, threw a cartridge from the magazine into the barrel, and raising it to my shoulder drew a fine sight and touched the delicate trigger. Instantly a sharp report rang out and resounded across the stretch of shingle and over the river which lay beyond. The great bird sprang suddenly upward with one convulsive start, then fell and lay quite dead alongside the lifeless body of the guanaco. When I arrived on the scene after leaving my place of concealment, neither the report of my rifle, the death of the condor, nor my presence among them had any but a temporary effect upon the carranchas. During my inspection of the condor after its death, these birds stood about at a distance of only a few feet. If I threw a stone or ran after them as though to frighten them away, they did not take to flight in a body, only those against which my efforts were more particularly directed would fly off for a short distance, or more likely make a short circle and return to their original positions.

During the three years that I spent in Patagonia I had abundant opportunities for observing and noting the manner in which the carranchas and condors attack the fresh carcasses of dead animals, and I always observed it to be similar to that described above. The carranchas always attack the eyes first, then cut out and carry away the tongue by making a hole through the skin at the base of and between the lower jaws, while the condors prefer the abdominal viscera and, in order the more readily to get at these delicacies, they attack the carcass in the region of the flank, where the external abdominal wall is thinnest and protected by very thin skin and a rather scanty covering of hair.

Contented with my slaughter for the day, I picked up the dead condor, stuffed some cotton in the wounds to prevent the blood from soiling the feathers and, taking my Winchester, started for camp with my heavy burden. After tramping some distance through the low sand hills that lay between me and the tent, I came upon Mr. Peterson busily engaged in the skinning of one of the three guanacos which he had killed from a bunch shortly after leaving me on the bed of shingle. Depositing my load upon the ground, I assisted him in his work and when the skins of the three were off and the flesh stripped from the skulls and limb bones necessary for properly mounting the stuffed skins, a horse was procured and all packed safely into camp, where they were treated with preservatives and properly cared for until in a condition fit to pack.

While we had gone to Patagonia with the mutual understanding that Mr. Peterson should devote himself especially to the recent fauna and more particularly to the birds and mammals, while I would look after the geology and palæontology of the region, yet there was in reality no division of labor. We both worked together, each giving his time either wholly or in part to the special work of the other, according as we were in a locality especially favorable for one or the other branches of our work. Palæontology, however, took the preference over all other branches, as was quite natural, since that was the primary purpose of ·our expedition to Patagonia, and most of our previous work in behalf of natural history had been in connection with vertebrate palæontology.

After several days spent in much the same manner and with success very similar to that of the day just described, I decided to spend a portion of my time in exploring the bluffs of the sea to the northward from Cape Fairweather in search of fossils. For some days I searched these beds, but with only very indifferent success. While walking along the beach on the morning of a particularly fine winter's day, for Patagonia, in July, while we were still camped at Cape Fairweather, I noticed a great block of hard sandstone weighing several tons lying at the base of the cliff, half buried in sand and literally covered with the fossil shells of gasteropods, brachiopods, giant oysters and other marine invertebrates. I was not a little astonished at this, for I had carefully noted the character of the rocks composing the bluffs as I walked along and had determined them to belong to the Santa Cruzian formation and of fresh-water origin, and no marine beds had ever been reported as overlying these deposits. From whence then came this great mass of sandstone, so clearly of marine origin? Its very size, as well as its sharp angles and unpolished surfaces, precluded its having been transported from a distance. On looking about I saw another smaller but similar block only a few feet above my head protuding from the talus which had here accumulated at the foot of the cliff. From this I traced the origin of this marine deposit still higher, until, at an elevation of some two hundred feet, I gained the top of a great land slide, where a tract about half a mile in length and from one to two hundred yards in width had broken away from the main mass and slid off into the sea, carrying with it a section of the plain above equal to the dimensions just given. In its slide downward, the bottom of the mass had evidently moved a much greater distance seaward than the top, so that the latter was left behind, as it were, and thrown backward, causing the different strata and what had once been the surface of the plain to be inclined at a high angle dipping downward and backward toward the face of the cliff, while the upper surface of the slide presented a very broken and uneven appearance, bearing abundant evidence as to the great disturbance which had taken place in its strata at the time of its displacement. Where not covered with talus, which had accumulated in great quantities at the foot of the cliff that still towered two hundred and fifty feet above, I detected in the upturned edges of this landslide a stratum

FIGURES 3 AND 4—SEE OTHER SIDE

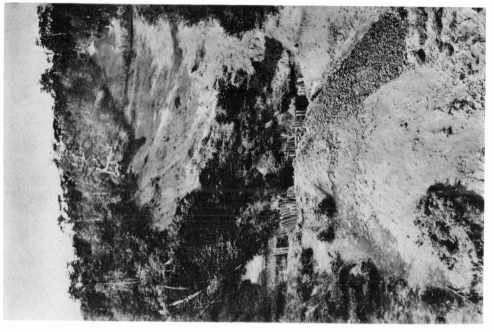

4- COAL MINES AT SANDY POINT.

3- CAPE FAIRWEATHER BEDS.

of marine sandstones some thirty feet in thickness, from which the block first observed lying on the beach had been derived. It was impossible to determine the exact stratigraphic position of the marine deposit from an examination of the top of the landslide. On turning my attention to the cliff above me, however, numerous large oyster shells appeared mingled with the sands and clays of the talus, while lying at the extreme top of the cliff, but beneath the shingle, the thirty feet of marine sandstones were clearly visible, as shown in Fig. 3. Here was a new discovery full of interest and meaning. The presence of these marine deposits overlying the several hundred feet of fresh-water and æolian deposits constituting the Santa Cruz beds, means that subsequent to the times when *Nesodon, Astrapotherium, Icochilus, Theosodon, Diadiaphorus, Boryhyæna, Phororhacos,* and numerous other genera of animals lived about the borders and left their bones to be buried in the mud then forming at the bottom and along the shores of the prehistoric streams, rivers and lakes of the Santa Cruzian epoch, this entire region had been buried beneath the ocean for the time during which these marine beds were deposited. I subsequently observed these new marine deposits, which I have called the Cape Fairweather beds, at various other and widely separated localities, and there is little doubt that they at one time covered all southern Patagonia, though at present they are wanting over large areas, having been removed by erosion. In the foothills of the Andes I found them capping the mountains at altitudes of five thousand feet or more, thus demonstrating what an enormous movement the crust of the earth has undergone over this region since their deposition in the Pliocene seas.

The top of this landslide, from which was made the interesting discovery detailed above, was capital collecting ground. Not only were we able to secure a very complete series of the marine invertebrates belonging in the beds at the summit from the blocks of matrix that had tumbled down, but the narrow and irregular valley which lay between the face of the cliff and the front of the landslide was covered over with calafate, mate verde and mate negra bushes which, on account of the peculiar protection the location afforded them, attained a size considerably exceeding that of the same bushes on the wind-swept plain above. For this reason this cozy corner was the favorite haunt of many birds and small mammals. In the midst of the dense foliage of the *mate negra* the little brown wren, *Troglodytes hornensis*, sends forth a perpetual chirrup. Constantly on the move, he hops from branch to branch, the brown tail, ornamented with delicate cross-bars of a darker shade, standing erect. If disturbed the head is suddenly elevated and each feather of its delicate crest thrown on end, as he violently stamps his feet, accompanying each movement with a series of sharp, discordant notes, which are hurled with great volume and rapidity from the delicate throat and heaving breast. At such moments the entire attitude of this tiny creature is defiant and is the cause of no little amusement to the intruder. Flocks of the red-breasted meadow lark, *Trupialis militaris*, may be seen flitting about among the larger bushes. *Phrygilus melanoderus*, the black-throated sparrow with yellow and lilac shoulders, is also common, while the chestnut-crowned song sparrow, *Zonotrichia canicapilla*, is omnipresent here, as elsewhere in Patagonia. Scarcely a moment throughout the day that its cheerful notes could not be heard. In storm or sunshine they would break into sweetest song, as though to dissipate a little the solitude and loneliness of the surroundings. Aside from these smaller birds, the projecting ledges of the overhanging cliffs, towering to a height of two hundred and fifty feet, were the favorite haunts of the condor and various large hawks and owls. The many cracks and crevices in the surface of the landslide were frequented by a host of small rodents, while the beautiful little gray fox or wild dog, *Canis azarœ*, was also plentiful here, as everywhere throughout the Patagonian plains. This little carnivore, while commonly called a fox, belongs more properly to the lupine or thoöid series of *Canidœ* than to the vulpine or alopecoid series. Not only is its structure wolf-like rather than fox-like, but its habits are also decidedly more similar to those of the wolves than of the foxes. It is both nocturnal and diurnal, not at all shy and easily approached. These animals are of an extremely playful and mischievous disposition, but without any of the cunning which, by common consent, has been ascribed to the foxes. At times their actions and deportment are not unlike those of a half-grown shepherd dog. They are extremely fond of rawhide or leather, and when by any chance articles made of either were left about camp within their reach for any length of time, such articles were sure to be found in an entirely ruined condition. This necessitated our placing everything beyond their reach when not in actual use. On one occasion I loaned my saddle and bridle and the borrower, on returning them in the evening, was not careful to place the latter in a safe place. As a consequence when I next wanted to use it, I found only the bit and buckles ; the reins and head pieces, which were made of California red leather, were cut up into bits each not more than an inch in length. The damage wrought by these little animals would seem to be due to an inherent spirit for wanton or mischievous destruction rather than a search for food. It was never safe to picket a horse with a rope made of rawhide or a long tie strap made of leather, since either might be found cut to bits by these animals.

They live in shallow burrows, among bushes and in the crevices of the rocks, where such are to be found. They seem to be chiefly scavengers, living for the most part upon the carcasses of dead sheep, guanaco, and other animals. They undoubtedly prey on smaller mammals and on the eggs and young of birds, when the latter are in season. Their fur is abundant, of a soft quality and rather light gray color over most of the body.

For some time succeeding my discovery of the Cape Fairweather Beds I used to go daily to the summit of this landslide, to attend a number of traps which I had set, not without success, for small mammals, and to col- lect fossils from the newly discovered marine horizon. On one of these days, while engaged in collecting a series of most beautifully preserved brachiopods from a block which had broken loose and fallen down from the summit and lay on the surface of the slide, a condor, apparently at- tracted by my presence, alighted on a projecting ledge of the cliff above me and sat for some moments apparently intent on ascertaining the cause of my presence in such an unexpected and inaccessible place. It had been my custom on such days to take with me a double-barreled shotgun together with a number of shells loaded with shot of various sizes. At that very moment this weapon lay at my feet. Already familiar with the habits of these birds I was in no hurry, but leisurely taking up the gun I slipped a couple of shells loaded with BB shot into the chambers, approached a few steps nearer and, just as the noble bird was preparing to be off, discharged the contents of one barrel. He dropped instantly, landing with a thud on the soft talus and lay stone dead (?) almost at my feet. Going nearer I picked up the dead bird and taking it with me I carefully laid it with my gun upon the ground and resumed the work of collecting brachiopods and other fossils, which had been thus temporarily interrupted. After continuing the latter for a number of hours, or until such time as the fading twilight indicated that the day was spent, I packed and placed the fossils safely in my collecting bag and, taking up the condor, started for camp. All went well until I reached the steep incline leading from the summit of the landslide to the beach two hundred feet beneath. On previous occasions I had found the descent of this a difficult matter and I hesitated to attempt it with the present additional encumbrance. While meditating as to which of the articles I had best leave, and loath to part with any, it suddenly occurred to me that, since the condor had not suffered from its fall when shot, it would be likely to receive little or no injury if I rolled it over the cliff to the beach below. Once on the beach I could take all to camp without further trouble. This seemed an easy solution of the difficulty, and I proceeded at once to act upon it. Placing my gun and other articles on the ground, I took the condor in both hands and threw him as far out as possible, intending that he should fall upon the shingle of the beach without striking any of the projecting ledges of the cliff. What was my surprise and disappointment, however, to see the bird almost immediately after he left my hands take to flight and go soaring off as though quite unharmed, bearing with him the highly prized skin with the beautiful coat of brilliantly white and jet black feathers which I had for several hours confidently considered as one among the number of the more important of my acquisitions for the day. Verily we should not count our chickens before they are hatched.

We remained at this camp near Cape Fairweather for more than a month, or throughout midwinter. Notwithstanding the generally raw and inclement weather, with occasional short spells of quite severe cold, I was surprised at the tenacity and hardiness of some of the wild flowers, and more especially one species of Compositæ, a *Senecio*. I found this plant in full bloom in an especially favored locality on the top of the landslide mentioned above, on the fourth day of July, 1896, a day equivalent in the northern hemisphere to the fourth day of January. This seems a little remarkable, considering that Cape Fairweather is located in S. Lat. 51° 30', or seven hundred miles farther from the equator than New York City, and notwithstanding that the winter of 1896 was an exceptionally mild one for that country.

Early in August we left Cape Fairweather, moving to a point (Cañon de Palo) some miles farther up the coast, where we remained for a few days, examining the bluffs for fossils and in making further collections of recent birds and mammals. Not meeting with very great success at this locality, after a week we moved on some ten miles farther north, where we spent the remaining portion of the month of August, collecting a considerable number of birds and mammals, but finding fossils still scarce. One day, when encamped at this place, while walking northward along the beach, scanning the bluffs in search of fossils, I came upon the body of a whale which had at some previous time been cast upon the beach and lay partially buried in the shingle. The thick hide was still intact, entirely covering the bony skeleton. Indeed as I saw it at this first visit the carcass seemed to have suffered little from decomposition. To satisfy my curiosity I paced its length, thirty-one steps as it lay extended on the shingle. The great body of this huge animal of many tons weight, as it lay weathering on the beach at an altitude only reached by the highest spring tides, furnished a striking example of the force of the tidal wave by which it had been borne in and left stranded in the position it then occupied. I looked longingly at it and wished with all my heart that I might include its skeleton in our collection of recent Mammalia, well knowing how much there still remained to be learned concerning the Cetacea of these southern seas. However, with the small means at my disposal and the personnel of the expedition limited to Mr. Peterson and myself, this thought had to be reluctantly dismissed.

CHAPTER IV.

Corriguen Aike; Character of the beach; Abundance of fossil bones and footprints in the rocks of the beach; Collecting fossils from beneath the sea; Dense fogs during month of September; The spring tides; Actions of the grebe in the surf; The sea leopard; Porpoises; A Patagonian spring; Spring flowers; Bird life in spring time; Habits of the Ibis; Spur-winged plover; The grouse-like plovers of Patagonia; Water fowl; Flamingoes; Upucerthia dumetoria; Lizards and frogs; Insects; Eggs of the Rhea; Second shipment of fossils.

ON the first of September we moved camp to Corriguen Aike, some twelve miles south of Coy Inlet. On almost the very first day after our arrival at this place we discovered a locality extremely rich in vertebrate fossils on the beach some two miles farther north. The different strata constituting the bluffs of the coast dip very gently to the southeast, and the beach, as it disappears beneath the waters of the Atlantic, is inclined at about the same angle as the beds. So gentle indeed is this incline that between the east coast of Patagonia and the Falkland Islands the sea nowhere attains a depth of more than one hundred fathoms. The gentle slope of the beach, together with the enormous rise and fall of the tides, amounting to over forty feet, results in a wide belt, extending seaward from the foot of the cliffs, which is entirely submerged by the rising tides and appears at low tide as a broad, shelving beach, its surface swept clean by the receding waves and elevated only a few feet above the surface of the water. In many places this beach at low tide extends seaward for a distance of two miles, or even more at some points along the coast.

It was in the sandstones of this shelving beach, near Corriguen Aike, that we discovered the rich deposit of fossil bones mentioned above. At this point, as at most places throughout this beach, erosion has taken place along the bedding planes, so that over considerable areas the surface of the beach represents essentially the same geological horizon. At this particular locality the dark green sandstones in which the bones were imbedded bore evidence of having been deposited over the flood plain of some stream or shallow lake.   On walking about over the surface at low tide, there could be seen the skulls and skeletons of those prehistoric beasts protruding from the rock in varying degrees of preservation.   At one point the skull and skeleton of *Nesodon* would appear, at another might be seen the limbs or perhaps the teeth of the giant *Astrapotherium* just protruding from the rock, while a little farther on a skull and jaws of the little *Icochilus* grinned curiously, as though delighted with the prospect of being thus awakened from its long and uneventful sleep.   On one hand, the muzzle of a skull of one of the larger carnivorous marsupials looked forth, with jaws fully extended and glistening teeth, the characteristic snarl of the living animal still clearly indicated, while at frequent intervals the carapace of a *Glyptodon* raised its highly sculptured shell, like a rounded dome set with miniature rosettes, just above the surface of the sandstones.   Throughout eighteen years spent almost constantly in collecting fossil vertebrates, during which time I have visited most of the more important localities of the western hemisphere, I have never seen anything to approach this locality near Corriguen Aike in the wealth of genera, species and individuals.   The bones of these animals were not the only records preserved of their former existence, for at certain places, on looking across the surface of the sandstone, one could see their fossil footprints.   At one locality especially favorable, where the erosion had evidently taken place along a single bedding plane over a considerable area, a series of tracks was seen extending uninterruptedly for a distance of about one hundred feet, making it quite possible to determine the exact stride of the animal.   The presence of these tracks is conclusive evidence that these animals roamed over, lived and died in this very region during the time when the sandstones and shales which contain them were being deposited, and precludes the possibility of the deposits having been laid down over the bottom of a great lake, or any other large and stable body of water.   The origin of these deposits will be fully discussed when we come to treat of the geology of the region.

As will readily appear from the preceding remarks, our work of collecting fossils was restricted each day to the period of low tide.   The principal fossil-bearing locality was limited to an area about one and a half miles in length, with an average breadth of perhaps three hundred yards.   As the water gradually disappeared each day from this area with the receding tide, we would commence the work of excavating the fossils, taking up each skull, limb or skeleton in a block of matrix and placing it in a conspicuous place where it could be easily seen. Each day we would follow close upon the heels of the receding tide and, working as rapidly as possible without endangering the bones, take up one skull or skeleton after another, until the turn of the tide and the waters came setting in again, when the work of excavating was abandoned, in order to convey the material already secured to the shore and place each specimen in a place of safety above high-water mark. At times, when delayed longer than we had anticipated in the excavation of a particular skeleton, or having been enticed by reason of an overpowering interest in our work to continue it longer than was prudent, this work of transporting the fossils to shore became a work of rescue in every sense of the term, frequently quite exciting, and on one or two occasions the question of rescue became a matter so entirely personal to ourselves as to become exceedingly disagreeable.

All our material safely stored upon the beach, the remainder of the day would be spent in trimming the superfluous matrix from the different specimens, hardening the softer bones and otherwise preparing them to withstand the vicissitudes of their long journey from the Straits of Magellan to Princeton. The bones and skulls, as well as the matrix surrounding them, were of course thoroughly saturated with sea-water, and in that cold and damp climate several days were necessary before they were sufficiently dried to permit of packing for shipment.

Ours was a remarkable and interesting experience, as patiently pursuing our work, we sat on the surface of the talus-covered slope at a safe distance from the waters that dashed furiously beneath, with the enormous wall at our backs, rising perpendicularly to a height of more than four hundred feet, while over the sandstones of the beach, from which but a few hours previously we had been excavating the remains of prehistoric animals, there now rolled a sea sufficiently deep for the safe navigation of the largest transatlantic liner. What a remarkable change, and in so short a time ! Not many experiences even among those of my childhood, that most impressionable period of our lives, have left themselves so indelibly engraved upon my mind. At this somewhat distant perspective, when not engaged in any particular line of thought, as sometimes happens, I frequently detect myself in the midst of a most vivid mental picture drawn from some particularly interesting point along this coast.

Here seated in the foreground on some convenient ledge intently listening to the deep sonorous music of the waters beneath, as the crest of each great tidal wave comes rolling in from its long journey across the eastern seas and breaks with thundering force upon the rocks below, once again I am for a moment transported to some favorite spot and busily engaged in my chosen work, only to awaken almost immediately from my reverie to a painful consciousness of the delusion and to wonder whether or not the desire was father to the dream.

All through the months of September and October we worked at this locality. There was a convenient and safe place for our horses at the back of a little landslide near the foot of the cliff adjacent to our work, so that we could ride down in the morning and back at night. The month of September was in many ways a peculiar one. Not only was it early spring with its equinoctial storms and high spring tides, but with us it was especially characterized by the the dense fogs which hung almost continually over all, so completely enveloping everything that it was impossible to see more than a few yards, or rods at most. Frequently during this month, while working on the beach at low tide, for days at a time the fog would be so dense as to completely obscure the bluffs of the shore, and where there were no rivulets to direct our course we had to resort on such occasions to the expedient of giving a sharp hallo and then direct our steps in that direction from which the echo came if we wished to reach the bluffs. On such days, though separated from one another frequently by a half mile or even greater distances, we could hear distinctly each other's pick strokes and could even converse without much difficulty. If, perchance, one of us wanted some tool or other article in the possession of the other he directed his steps by the direction from which came the sounds of the pick strokes of the latter, and if, as sometimes happened, these were temporarily suspended, we were forced to resort to the peculiar procedure of enquiring of the whereabouts of one another. For twenty-eight days during this month we never saw the sun, save for a few moments in the early morning, and then only on three or four occasions. At such times, however, it would appear only for a brief interval and shine with a brilliancy unequalled by the early morning sun of this region at other times, giving promise of a bright and cheerful day, which would be almost immediately dissipated by the appearance of the ever-present fog. It was in this continuous fog that the sailing ship Columbia, bound round the Horn from England to Seattle, was driven ashore and wrecked on the coast just north of Port Desire. We saw Captain Bull and his crew of thirty-two men at Gallegos after they had been rescued by the Argentine transport Villarino, the same by which we had come from Buenos Aires to Gallegos. The officer jocosely remarked that he had been round the Horn many times and had always wanted to see the coast of Patagonia, and at last that wish had been realized, though apparently not in the manner he desired, for in conversation with me during a more serious moment he expressed a fear that his certificate might be taken from him. I never learned what disposition was made of his case, but I trust he received no censure and speedily received another vessel. For, from the difficulty we experienced with our work throughout the month of September, I can readily appreciate the great mental strain he was subjected to during all those days of almost total darkness, prior to the time when his good ship foundered on the bleak and cheerless coast near Port Desire.

During the period of the vernal equinox, when the tides run exceptionally high, the wind, as during most of this September, was unusually calm for Patagonia. In the early morning before the fog settled down over all, as we rode along the beach and looked eastward over the great ocean, it presented a surface like that of polished silver thrown into a series of long, parallel waves. On such occasions the average height of the crest of each wave, as it came rolling in upon the shore, was three or four feet. At frequent intervals, however, an especially high swell could be seen forming far out at sea. As this moved landward its volume and speed would be augmented and accelerated. So long as the depth of the water continued sufficient to prevent friction against the bottom it would move rapidly forward with a perfectly even and unbroken front. When, however, the shallower areas near the shelving beach were reached, the friction at the bottom would arrest the movement of the lowermost stratum of water, while that of the uppermost would continue uninterrupted. Thus, as the water in the rear came rushing on, there would result a piling up of the water in front, until the crest became actually overhanging and raised to an altitude of eight, ten, or even twelve feet above the base. Suddenly the crest would break and the great volume of water would plunge forward and rush madly up the beach with a deep, almost deafening roar, its irritated surface covered with a glistening mass of foam and spray.

On such mornings it was interesting to observe the movements of the different water fowl, which were everywhere abundant on this coast, and note the attitude of each toward the tide. I noticed that when the tide was highest and the force of the waves against the shore greatest, the gulls, cape pigeons, etc., remained far out to seaward beyond the limits of the breakers, where they sat in perfect safety, gracefully riding the successive swells as they came rolling in.   The grebe, on the other hand, seemed fairly to revel in the excitement of the surf and breakers nearer shore.   They defied the dangers of the smaller waves and breakers and rode in triumph through the dashing spray of each.   When confronted by one of the larger breakers just described, they resorted to other and quite different tactics.   On such occasions, when within a few feet of the high and often overhanging wall of water, which came rushing forward with maddening speed, accompanied by such terrifying and ominous sounds, just as the last hope for their safety was vanishing and their destruction seemed inevitable, they would suddenly dive, and, disappearing for a few moments beneath the advancing wave, would shortly reappear quite as suddenly, sitting gracefully and unconcerned upon the surface in the rear of the wave, awaiting with apparent eagerness an opportunity for repeating the operation.   They were not always successful, however, in the safe performance of this feat, as was clearly evident from the great numbers of dead and crippled birds found scattered along the beach.   It will readily appear how, by the miscalculation of only a few seconds, this interesting sport might end in the destruction of the participant.

On quiet days, when the tides were running at their highest, the waters immediately fronting the shingle-covered beach were frequented by considerable numbers of *Leptonyx weddeli*, the common haired seal or sea leopard of this region.   Occasionally these animals would approach quite near the beach, just beneath where we were engaged with our fossils, and thrusting their heads far out of the water remain stationary for a moment, apparently intent on ascertaining the meaning of our presence.

On several occasions when the tide was full, we were visited by schools of porpoises and I never tired of watching these graceful cetaceans disport themselves in the water about me.   At times they would move along with the great dorsal fins protruding from the surface of the water.   Then instantly, as though one mind controlled the movements of all, they would turn on their sides, or with back down glide gracefully along just beneath the surface of the water, the brilliant white of the belly and throat shading off into the delicate lilac grays of the sides.

We continued our work near Corriguen Aike all through September and October. With the advancing spring came important changes in the fauna and flora of the region, so that our interest in our surroundings increased rather than diminished, although at no time had it been lukewarm.

The change of seasons in a semi-arid country is never so radical as in one where there is an abundant rainfall and luxuriant vegetation. This is especially true of the semi-arid, treeless plains of southern Patagonia. Throughout the year the surface is clothed with but a scanty covering of brown and withered grass, which, even in springtime, makes only a feeble attempt at being green. The few flowers are, for the most part, either colorless, or small and inconspicuous. Some, however, like the delicate little pink oxalis, *O. laciniata*, one of the earliest harbingers of spring, two species of *Calceolaria*, one or two species belonging to the Liliaceæ, a yellow *Œnothera*, and a few others are of exceptional beauty.

While the bird fauna is pretty much the same throughout the year, there are a few migrants among the land birds, and in springtime, as the breeding period approaches, the new life awakened, as it were, in the avifauna is perhaps the most notable of all the features by which the change of season is announced. The ibis, *Theristicus melanopsis*, the presence of which we had not noted during the winter months, appeared in great numbers with the advancing spring. The high bluffs of the sea were at night a favorite haunt of these birds, and often when returning to camp from our work late in the evening, great numbers of them could be seen perched in convenient places about the cliffs, or flying about in and out among the crevices and ledges, accompanying their movements with those peculiar and irritating squawks and screams for which these birds are noted. A high cliff just in front of our tent was chosen as a roosting, and later perhaps nesting, place by a colony of these birds and the disturbance and annoyance they caused us was, at times, to say the least, most distracting. Up at daylight and off to our work, after toiling all day long on the cold, damp and perhaps wind-swept beach, we would return late at evening and, after partaking of a hastily prepared meal, retire for the night, only to have our slumbers disturbed and our much needed rest interrupted by the harsh cries and screams of these birds, often prolonged, though spasmodically, far into the night. If their noise had been continuous, we should doubtless have soon been able to accustom ourselves to it, notwithstanding its irritating nature, and thus find a certain relief. But so far from being continuous, it was quite the opposite. After a period of extreme quiet, during which not a sound could be heard, suddenly, as though by a preconcerted signal, they would pour forth such a pandemonium of discordant screams as would literally fill the air and rebound in every direction from the walls of the cañon about us. After keeping up this commotion and making night hideous for a period of perhaps four or five minutes, they would gradually settle back into a state of absolute quiet, which seemed all the more absolute by contrast with the extreme turbulence of the one just closed. This quiet would continue for from twenty to thirty minutes, perhaps, when we would be suddenly and violently startled from the peaceful slumber into which, through sheer fatigue, we had fallen, by the renewal, with increased energy, of this pandemonium, which would be poured forth with such volume as would lead one to imagine that all the demons of Hades had been suddenly let loose to wreak vengeance upon us for our desecration (?) of mother earth's ancient sepulchres. The apparent propensity shown by these birds for passing a considerable portion of the night in riotous disturbance was a source of annoyance from which, during the month of October, we could find no relief. They were particularly annoying to my companion, though I cannot lay claim to having derived any pleasure from being an unwilling listener to their nocturnal concerts. Indeed, the sensations were similar, though augmented by many diameters, to those more familiar which one experiences when tying to snatch a few hours of much needed rest, while two representatives of the Felidæ, perched conveniently near the windows, with voices attuned to concert pitch, indulge in one of those spirited nocturnal concerts for which tabby is so justly famed. For a time the annoyance we experienced from these birds was such that we resorted to all sorts of devices to rid ourselves of them. On several occasions Mr. Peterson was so exasperated that he took the shotgun and fired several volleys into them. This had only a temporary effect, however, for while he would succeed in dislodging and driving them away, they would never fail to return shortly afterward to their accustomed place. But with all his imperfections the ibis has some good qualities. While they were an undoubted source of annoyance to us throughout a considerable portion of the night, shortly after midnight they would settle down to a state of perfect quiet, which would continue uninterrupted until the early morning, when, with considerable noise and confusion, they would sally forth from their eyries in the cliffs to feed during the day on the broad level pampas. In this they were as regular as an alarm clock, and as they went screaming over our tent toward their favorite feeding grounds. on the higher pampas, we knew that it was time to get up and prepare for the day's work. On returning to our work on any morning, the scene of so much activity during the previous and succeeding evenings would be entirely deserted. It was evident, therefore, that for the present at least these cliffs were used by these birds only as roosting places. I was assured, however, by the natives that later in the season many of them might be seen nesting in the cliffs. Concerning the accuracy of this statement I can say nothing, never having either verified or disproved it by my own observations.

Another bird which made its appearance with the early spring was the spur-winged plover, *Belonopterus chilensis*, a rather handsome bird, prettily marked with white and black, with a long spur of a delicate pink color protruding from the wrist of either wing. These birds usually go in droves of from ten to a dozen and frequent the margins of shallow ponds and lakes, where they stalk about on their slender, but not over long legs of a rich pinkish tint, and feed on the tender shoots of grass and, no doubt, certain species of insects as well. Like the ibis, they are extremely noisy birds, especially when disturbed either by night or day. While on the ground and undisturbed I do not recall that they were less quiet than other birds of their class. On taking to flight, however, whether of their own volition or from being disturbed by an intruder, they accompany such operation by a series of loud shrill notes uttered in rapid succession, which are continued at intervals throughout their flight, although they suddenly relapse into absolute quiet on alighting again on the margin of the same or an adjacent pool. Judging from the frequent flights which I have observed these birds to make, apparently at all hours throughout the night, I should say that they were nocturnal, at least to the extent of being indifferent as to any particular time of day or night for feeding purposes. Freqently throughout my subsequent travels in Patagonia, when encamped convenient to a lagoon frequented by these birds, and not particularly desirous of sleep, I have lain awake throughout the greater portion of the night, intently listening to the notes and flight of these interesting birds. Every few moments they could be heard in rapid flight, accompanied by their peculiarly shrill notes, as they shifted their position to other feeding grounds about the shores of the lake.

A small, gray, grouse-like plover, *Thinocorys rumicivorus*, somewhat larger than the English sparrow, with white throat, and breast marked by a deep black cross, was abundant on the higher pampas, where coveys numbering from fifteen to twenty individuals were not uncommon. Several other species of larger grouse-like plover were to be found, though somewhat less abundantly, over the pampas. Among these may be mentioned *Thinocorys orbignianus*, the smaller white-bellied plover; *Attagis malouinus*, the larger white-bellied plover; *Oreophilus ruficollis*, the black-bellied plover, and less common and in more sheltered places specimens of *Attagis gayi* were to be seen. The latter is of about the size of our common sharp-tailed grouse and with habits somewhat resembling those of that bird. All the different representatives of this family of birds in Patagonia have very similar habits, resembling more nearly those of the grouse than of the true plovers. As a rule they are all exceptionally shy, though depending quite as much for protection upon the coloring of the feathers of their backs and wings, which are singularly well adapted for protective purposes, as upon their power of flight, which, though rapid, they are incapable of sustaining for any considerable distance. When approached while feeding on the pampa, they nestle down and, with head and neck extended, lie perfectly prone upon the surface of the shingle-covered plain. Their color harmonizes so well with that of their surroundings, when in this position, that they are extremely difficult to detect. I have seldom seen, among birds at least, a more perfect or effective example of protection afforded through adaptive coloration. On account of the delicious quality of their flesh these birds are highly prized as food. Frequently when in quest of such sport, having detected a covey of plover quietly feeding on the almost barren surface of the pampa, but at a distance quite beyond the effective range of my fowling piece, I have observed that, although they were clearly visible when first detected and remained thus until such time as their attention was attracted by my approach, they would then suddenly and mysteriously disappear, as if by magic. Knowing that they had not taken to flight, I would approach the spot where they had last been seen, confident of regaining sight of them as I came more closely. As a rule my expectations were disappointed, and I would arrive upon the scene only to find all deserted. Upon making a circle of from fifty to one hundred yards, however, I would usually come upon them, though so completely obscured, as they lay at intervals of only a few feet, hidden in shallow depressions or concealed behind the scanty tufts of grass, that I would only become aware of their presence by the startled flight of one of their number, as it rose suddenly and directly from my feet, where it had lain and might have remained unnoticed and in perfect security, unless accidentally trampled upon. On remaining perfectly quiet for a few moments, after thus disturbing one of these birds, and carefully scanning each tuft of grass and the inequalities in the surface in the immediate vicinity, one may detect the other members of the covey, each bird lying flat and perfectly motionless like so many inanimate objects on the shingle-covered surface. I have said "like so many inanimate objects," though as a matter of fact this is not strictly true, for upon closer inspection two small, jet-black eyes may be seen intently watching the movements of the intruder, and immediately upon detecting any intention in the latter hostile to themselves, they are off in an instant, accompanying their rapid and peculiarly spiral or zigzag flight with a series of low, plaintive chirrups or notes not at all unpleasing to the ear and calculated to soften the heart of the sportsman, if indeed he has not already forgotten the presence of his fowling piece through his greater interest in the novelty of the situation. The lives of thousands of these birds must annually be saved through the protection afforded them by color, from destruction by the numerous hawks, falcons, and other predatory birds and mammals that throughout the year infest this region in great numbers.

A considerable variety of ducks, geese and other water fowl were at this season especially plentiful about the margins or in the waters of the rivers and lakes. Scarcely a pool or marsh, however limited in area, that was not tenanted by at least one pair of the American grebe, or "hell diver," *Podicipes americanus*. On the surface of the larger lakes there floated gracefully great numbers of the black-necked swan, *Cygnus melanocoryphus*, while the beautiful white-necked variety, *Coscoroba coscoroba*, though much less common, was occasionally to be seen clothed with a covering of immaculate white feathers, either solitary, or with one or more companions, gliding in and out among its more numerous relatives. Stalking about in the shallower waters of the lake were groups of flamingoes, *Phœnicopterus chilensis*, busily engaged with their long bills in extracting from the mud at the bottom the worms, crustaceans, molluscs, and such other animal life as would serve for their sustenance. During this exceptionally open winter we observed that occasional examples of these birds were to be seen throughout the year, even at this high latitude. Although quite common throughout the most of the year, about the lakes all over the Patagonian plains, I was never able to discover a nesting place of these birds, though constantly on the lookout for such. I was told, however, that there is a lake in northern Tierra del Fuego where in summer they congregate and breed in great numbers. The beautiful color of these birds is well known. It is quite variable, the red running through every shade from a light delicate pink in the younger females to a deep flaming crimson with the fully adult male. The black of the wings is intensified or diminished in the different individuals, according as the red of the other parts of the body is of a deeper or lighter hue.

Among the smaller birds beside the song sparrow, *Zonotrichia canicapilla*, black-throated sparrow, *Phrygilus melanoderus*, the little brown wren, *Troglodytes hornensis*, and a few others mentioned above, there may be mentioned as of more than ordinary interest *Upucerthia dumetoria*, a small brown bird with long curved bill. This bird is somewhat smaller than our meadow lark and has a peculiar and amusing habit of rising suddenly to a height of from fifty to sixty feet above the surface of the ground. Here for a few moments it will sustain itself in practically the same position by a rapid vertical movement of the wings, then, suddenly stopping, dart swiftly and almost directly downward until within a foot or two of the earth, when it will again as suddenly change its course and, gliding upward at an incredible speed, resume its former position in mid-air, repeating the same manœuvres over and over again for hours at a time, breaking forth all the while in a not unpleasing song, the intense volume of which is only decreased during the short interval of its rapid descent and ascent. The entire aspect of the bird while thus engaged is such as would indicate the keenest enjoyment. This same bird has the further peculiarity of building its nest in a shallow burrow. I frequently saw it nesting and observed that the nests were as a rule placed on a hillside, or near the top in the side of a small draw. The site for the nest would be chosen at the base of a little escarpment only four or five inches in height and most usually directly beneath a tuft of grass growing on the crest of the latter.   At the foot of the tiny escarpment a little tunnel three or four inches in length would be driven back into the side of the hill and at the end of this was placed the nest.

The little black *Scytalopus magellanicus* was abundant at this season of the year, both along the beach and over the pampas, where they moved about in considerable flocks, displaying habits not unlike those of the common shore lark, *Otocoris alpestris arenicola*, of our western plains, though not at all resembling the latter in color.

As the spring advanced, there appeared about the cliffs and over the pampas a great variety of small lizards of varying size, shape and color, but no snakes.   Throughout the three years spent in southern Patagonia I never saw or heard of a snake having been observed there.   Frogs were fairly abundant about the springs and pools of fresh water.   Beetles were abundant everywhere on the pampas, as were also spiders and a large scorpion.   There were butterflies and moths, though in no great abundance, and such as were present belonged for the most part to small and inconspicuous forms.

Two animals, the guanaco and the rhea, play such an important part in the economy of the aborigines of this region and indeed of the traveller throughout the interior that I shall leave the discussion of their peculiarities for the chapter treating of the Indians of Patagonia.   I must not omit to mention, however, in this connection that the eggs of the ostrich, *Rhea darwini*, provided a welcome and palatable addition to our daily menu. Throughout the season from October first until the last of November we had them almost daily.   Fried, scrambled, roasted in a bed of coals or mixed with flour and made into batter and baked as cakes, they were always relished.

With the end of October we had finished our work at Corriguen Aike and most gratifying had been our success.   In all we had collected at this locality alone about four tons of fossils, and as we unloaded the last of the boxes containing these on the beach at North Gallegos in the wool shed belonging to Mr. Felton, who had generously granted us free storage, a small schooner, "La Patria," of about forty tons, came in and anchored in order to discharge some cargo.   As she was going direct to Sandy Point, this seemed an excellent opportunity for making a second shipment. I was not long in striking a bargain with her Portuguese master, who was so illiterate that he had to call in the mate to compute the charges on four and one half tons of cargo at fifteen pesos per ton. He was a very good and honest man, however, and after learning that I was somewhat acquainted at New Haven and New London, Connecticut, where he had himself spent nine years, he was extremely obliging. On account of certain laws prohibiting the exportation of fossils from Argentine territory we were not at all displeased with this opportunity of starting this second consignment on its way home. The boxes were soon on board La Patria and stored safely in her hold, having been consigned to Braun and Blanchard in Sandy Point, with instructions to forward them by the first steamer to New York.

CHAPTER V.

Camp at Coy Inlet; By horseback to Sandy Point; Rio Chico; An accident; Ooshi Aike; Posada de la Reina; Cabeza del Mar; Cape Negro; Sandy Point; Return to Gallegos; Make preparations for a trip into the interior.

HAVING completed our work at Corriguen Aike, on November second we moved camp to the point of land directly south of Coy Inlet at the mouth of Coy River. Considering the importance of the shipment we had just made by the "Patria" and that we were quite unknown to the parties to whom we had consigned them in Sandy Point, we decided that it was best I should go myself to Sandy Point and personally look after their reshipment to New York. I was further influenced in favor of this decision by the hope that there might be letters from home awaiting us at that port, for though we had now been absent more than nine months and had written regularly giving full instructions as to how we were to be addressed, not one line had either of us received from those at home in whose health and welfare we were most deeply concerned. Accordingly on the following morning, November third, leaving Mr. Peterson to continue the work at our new camp, I set out on horseback for Sandy Point, distant some two hundred and twenty-five miles on the northern shore of the Straits of Magellan and about midway between Capes Virgin and Pillar. I succeeded the first night in reaching Killik Aike, where I found Mr. Felton and family at home, and experienced for the first time their generous hospitality. They were astounded when I told them I was on my road to Sandy Point, and assured me that with one horse I would never be able to reach my destination, even going so far as to offer me the loan of another. Accustomed in our own country to making trips of from five hundred to a thousand miles with one horse, I felt no alarm at undertaking so insignificant a journey as this seemed, more especially as I had been assured that at no point along the way was there a greater distance than fifty-five miles between settlements. At any rate, I did not feel the necessity of another horse and thankfully declined my host's most generous offer. The following morning Mr. Felton very thoughtfully provided me with letters of introduction to the various estancieros whom I would likely fall in with along the route, and accompanying me to the crest of the high bluff overlooking the Gallegos River, carefully pointed out the various landmarks visible from our position and indicated the general direction and principal features of the trail with such precision that I had no further difficulty on that score.

Bidding my host good bye, I started at a moderate gait for Guer Aike, determined, if possible, to reach the estancia of Señor Kark on the Rio Chico, some ten miles below Palli Aike, where I had been assured by Mr. Felton that I should meet with a most hospitable reception and could pass the night in comfort. In this I was not disappointed. Not only was I most hospitably received and generously entertained by both Mr. Kark and his wife, but on resuming my journey the following morning, that most amiable lady insisted on providing me with a lunch, assuring me that I should need it, in order to withstand the fatigue of the sixty-five-mile ride which lay between me and Ooshi Aike, which I should be compelled to make that night, or sleep on the high, barren pampa lying between Palli Aike and that place. With some reluctance I took the lunch so thoughtfully prepared by my hostess, little thinking at the time with what appreciation it was destined to be devoured. Thanking the lady for her kindness and accompanied by Mr. Kark, who had volunteered to go with me for a short distance, in order to point out the proper place at which to cross the River Chico and to indicate to me a cut-off by which I could save some three or four miles. His mission fulfilled, he bade me good bye and returned, while I continued on my journey. At this point the River Chico meanders through a narrow valley enclosed on either side by low hills capped with basalt. The river flows very near the surface, with swamps and marshes on every side. There is an abundant growth of tall grass which affords many advantages favorable for the nesting of the myriads of ducks, geese, and other water fowl that frequent this stream. After travelling a few miles along the south bank of the river, I came to the tongue of a comparatively recent lava stream, which had been ejected from an old volcano lying a few miles to the south of the stream and had flowed down over the sides of the bluff and lay spread out as a thin sheet extending for several rods over the surface of the valley.   Shortly after passing this I came to the cut-off pointed out by Mr. Kark, and, leaving the main trail, struck off to my left through the hills to the high pampa beyond.   After travelling some miles across the pampa in the general direction indicated by my guide, I again fell in with the trail, which, however, was not so plain as it had been.   As I continued my journey the altitude of the plain rapidly increased and the character of the country changed.   Scattered about over the surface were immense angular blocks of granite or other crystalline rocks which must have come from the Andes and could only have been transported to their present resting places through the agency of ice.   As I rode along there were frequent patches of half-melted snow, while pools of water caused by the latter lay about on every side.   The sky was overcast with leaden clouds and a fierce southwesterly wind beat savagely in my face, greatly retarding the progress of my horse.   Halting for a moment at one of these .pools, I dismounted in order to stretch my limbs, while my horse grazed on the scanty grass and refreshed himself with a drink from the pool.

On returning to remount my horse and resume my journey, I met with an accident, which, if it had happened under more favorable circumstances, would scarcely have proved in any way serious.   On dismounting I had thrown the rein over the horse's head, as is the custom in the western part of our own country.   When I returned I found he had one foot in the rein, and, reaching down to take it out, as I had frequently done on similar previous occasions, for some reason he became startled and jerked his head violently upward until the reins became taunt, when it was thrown downward again with even greater violence just at the moment when I was rising from the stooping position I had assumed, and in such a manner as to strike my head with the broken shank of the Logan bit with which the bridle was fitted.   This was forced through and under the scalp in such manner as to loosen the latter over a considerable area, at the same time rupturing some of the blood vessels and causing the wound to bleed very profusely.   For an hour or more I tried vainly to staunch the wound by bathing it in cold water, which was the only remedy at my disposal.   Not being successful in this I placed a handkerchief over the wound and resumed my journey, thinking that the flow of blood would soon stop of its own accord.   After travelling for some distance with no apparent cessation of the bleeding, with my upper clothing already saturated with blood, and a feeling of faintness gradually creeping over me, as I came to a wide, shallow basin where the grass was unusually long and thick, I decided that prudence dictated that I should stop, since the constant motion of my horse was evidently not beneficial to my wound. No stone or bush being near, I picketed my horse with one of the long bones from a guanaco skeleton and, cutting a few handfuls of grass, which I put in a convenient place, I unsaddled and, placing two pocket handkerchiefs over the wound, drew my Stetson hat down tightly over them, and wrapped in saddle blankets and slicker, with my saddle for a pillow, I lay down on the ground, protected from the dampness beneath by the few handfuls of grass previously prepared. I confidently expected that I should be able to resume my journey the same day and to reach Ooshii Aike late that night. However, such was not to be the case, and while the wound did not bleed so profusely, it was far into the night before it ceased altogether, and I finally fell asleep.

When I awoke the following morning, weary, weak and chilled to the marrow by the cold and penetrating wind, I saddled my horse and resumed my journey. I arrived at Ooshii Aike about ten o'clock the same morning, cold, hungry, and well-nigh exhausted, for I had had nothing to eat except the lunch given me by Mrs. Kark. Instead of meeting with the warm and hospitable reception I was so much in need of, I experienced the only rude treatment I ever met with throughout all my travels in Patagonia. When I applied for admission at the door, I was met by a surly Italian who, as I afterwards learned, was officiating as cook. To a request for something to eat and the use of soap and warm water with which to wash and bathe my wound, he returned a negative reply. My weakened and altogether miserable condition was plainly visible and would under ordinary circumstances have bespoken for me at least humane treatment. In vain I told him of my protracted fast, the night on the pampa, and the painful accident with which I had met. I made it clear to him that I was quite willing to pay for any accommodation he would grant me, and lest he might think I was deceiving him, displayed to him my ability to do so. He remained obdurate. As I stood for an instant considering what I should do, well knowing I was unfit to attempt the additional twenty-five miles that lay between this place and the next settlement, I decided that, if there was anything which would minister to my immediate necessities within the house, this " dago " should not stand between me and my needs.   In an instant I declared my intentions, and shoving by him, walked through the long hall to the room in the rear which I had correctly judged to be the kitchen.   In a capacious cupboard occupying a corner of the room, there was an abundance of excellent bread and cold mutton, sufficient for the wants of a dozen hungry men, while in a range at the opposite end of the room there still smouldered remains of the fire by which the morning coffee had been prepared.   This only needed replenishing from the pile of dry calafate that lay convenient, to burst forth into flame, imparting such warmth and good cheer to the room as, through contrast with the previous night passed on the cold, wind-swept pampa, made it appear not only comfortable, but even luxurious.   Making myself quite at home, after the manner of the frontiersman, I quickly prepared a pot of coffee, meanwhile bathing my head with warm water, until I could safely remove my hat and handkerchiefs without further injury to the wound, which I then cleansed and dressed as best I could.   Having refreshed myself sufficiently from the supply of bread and meat and, thoroughly warmed and stimulated by the cheerful fire and delicious coffee, I resumed my journey, not neglecting, however, to leave my card with a brief note for the foreman written on the back thereof, informing that gentleman of my depredations, and that on my return in a few days from Sandy Point I hoped not only to make his acquaintance, but such restitution as should be thought necessary.

I determined if possible to reach that evening "La Posada de la Reina" (the Queen's Hotel), some forty-five miles distant, where there was a public hostelry presided over by a fellow countryman from Virginia, one Taylor by name, where, I had been assured, I should meet with a hospitable reception and be made fairly comfortable.   After leaving Ooshii Aike, the trail leads for a distance of some twenty-five miles across a broad, level pampa somewhat lower than that I had passed over since leaving the Rio Chico, but quite as bleak, to Dinnemarcara, an estancia owned by a Spaniard and situated on a small stream, just where it issues from a narrow gorge cut through the low anticline to the west, which forms the easternmost of the foot-hills of the lower Andes.   I stopped at this ranch long enough to permit my horse to feed and to partake myself of some refreshments, so that it was late in the afternoon when I started for the Queen's Hotel, some twenty miles farther on.   For a considerable distance the trail led along the plain at the foot of the bluff, which was interrupted here and there by a number of small streams, so that the country around was well watered and capable of supporting considerable vegetation. The grasses were taller and covered the ground much more completely than in the pampas to the north. Bushes were also more abundant and of larger size, and had not the day been an extremely disagreeable one, I have no doubt there would have been a noticeable increase in the abundance and variety of birds and other animal life. Throughout the day I had noticed that my faithful horse, which heretofore had needed little encouragement, was showing unmistakable signs of fatigue. This rapidly increased with each succeeding mile, so that as daylight disappeared, accompanied by a cold, driving rain which only intensified the darkness, I decided to camp for the night rather than urge my horse farther, or take any chances, on account of the darkness, of losing the trail, which, through being little used, was in places quite dim. Selecting a group of bushes with especial reference to the protection they would offer from the wind and rain, I unsaddled and put my horse out to graze. Collecting a bundle of firewood, I built a fire in a convenient place, and having properly adjusted my saddle, saddle blanket and slicker, lay down to pass a second not very comfortable night on the pampa.

I resumed my journey at an early hour on the succeeding morning. After travelling some distance through a series of low, rounded, grass-covered hills, I came to a small stream with deep narrow channel in the midst of an almost impassable marsh. It was clearly fortunate that I had not continued my journey the evening before. With some difficulty I succeeded in crossing this stream and reaching the summit of the low ridge on the opposite side, where I could see the hostelry at a distance of some five miles and at the farther edge of the broad, level valley that lay spread out at my feet. I was not long in reaching this place, where I remained some two or three days in order to nurse the severe cold I had contracted, as well as the wound on my head, which was now suppurating very badly. While here, I had an opportunity of despatching a letter to Mr. Peterson informing him of my accident and that it would be impossible for me to return as quickly as I had anticipated, so as to relieve him of any uneasiness.

On the afternoon of the second day after my arrival at this place I purchased a fresh horse, and, leaving the other to recuperate until my return, set out for Sandy Point, going as far as the hostelry of Mr. Macdonald at Cabeza del Mar that same evening. From Posada de la Reina to Cabeza del Mar the trail winds in and out among low, rounded hills, separated by small ponds and broad stretches of meadow lands. The general aspect of the whole is not unlike that of the sand-hills of western Nebraska. Cabeza del Mar (head of the sea) is the name given to the land-locked body of water which extends some fifteen miles inland from Peckett Harbor in the Magellan Strait. At the entrance it is quite narrow and easily forded at low tide, while inland it expands into a broad sheet of water several miles in diameter and of considerable depth. To the westward it approaches to within eight miles of Otway Water, a similar though larger body of water which extends inland from the Pacific. The narrow peninsula which separates these two bodies of water is but slightly elevated above the sea, so that on the following morning, shortly after leaving the hostelry at Cabeza del Mar, I gained a point of vantage where it was possible to see the waters of both oceans at the same time. After leaving Cabeza del Mar, the road leads for ten miles across a low level pampa to the mouth of a very small and unimportant stream rejoicing in the name of Fish River, though I am confident no fish ever ventured farther up this stream than the limits of the tide. From Fish River to Cape Negro is but a short distance, and here one gets the first glimpse of the wooded region of the lower Andes. Here, as elsewhere along these mountains, on the outskirts of the forests, the trees are small, scrubby, and irregularly formed. From Cape Negro to Sandy Point the road leads alternately along the beach of the strait, across a bit of meadow land, or through small forest-covered tracts, finally emerging, some two or three miles northeast of the town, upon the surface of the low, level valley which has been gradually extended farther and farther into the strait through the addition of material brought down from the adjacent mountains by the Rio de las Minas. I arrived at Sandy Point about noon of the tenth day of November, suffering with a very bad cold and with the wound on my head much inflamed and suppurating freely. Having found a hotel, which, like all others in this miserable place, as I discovered later, was extremely bad and uncomfortable, after seeing that my horse was properly cared for, I repaired to a physician to obtain such medical attendance as my case might need, in order that I might be able the earlier to start on my return journey. There were two physicians in Sandy Point, one of whom had been recommended to me. I confess that the impression made upon me by this gentleman at first sight was not a favorable one, and when a moment later he suggested bleeding as the initial treatment, I was not long in deciding that I would be my own physician and surgeon, well knowing that since the first night on the pampa after my accident I had been in no way suffering from an excess of blood. From an apothecary I procured some quinine, carbolized vaseline, absorbent cotton, bandages, and a few other simple remedies and returned to my hotel.

The day following my arrival in Sandy Point, having learned that our boxes were stored in a hulk in the harbor, if the open roadstead at this place can be called a harbor, I procured a boat and rowed out to see in what condition they had arrived thus far on their journey. What was my surprise and consternation, on reaching the deck of the hulk, to see the boxes lying about quite unprotected from the almost constant rains. The excuse given was that as they were to be forwarded by the first steamer for New York, it had not been thought worth while to lower them into the hold. I immediately had them all piled together and securely covered over with tarpaulins, of which there were several lying about the deck.

In spite of every care my wound healed slowly, and for a few days my cold, far from improving, seemed to grow worse, so that I was for a time confined to my room. As my condition improved later and I was able to get about the town, or make short excursions beyond its limits, I had ample opportunity for seeing the sights of this truly cosmopolitan place. At the time of my first visit, Sandy Point claimed a population of from ten to twelve thousand and actually had perhaps four or five thousand. It is built at the southwestern end of a small valley at the mouth of the River of the Mines mentioned above. For the most part, the buildings consist of small wooden or galvanized-iron structures, although there is a small and constantly increasing number of substantial houses built of brick and stone. It derives its chief importance from being the seat of government of the Chilian territory of Magellanes, a port of call for all steamers passing through the Straits of Magellan bound for New Zealand, or the west coast of America, and the principal distributing and shipping point for goods received from and destined to Europe, for the already not inconsiderable and rapidly increasing wool-growing industry of Patagonia and Tierra del Fuego. Lumbering, in a small way, is carried on in the adjacent forests. This industry could be increased many fold. The town itself is well laid out, but very indifferently kept. Sidewalks are rare and street-crossings, where there are any at all, consist of logs laid end to end, so that when the streets are muddy, as is generally the case, one must needs be an experienced tight-rope walker to get about in safety. None of the streets are paved, and as for a sewerage system, not even an attempt is made at surface drainage, although the topography is such as would greatly facilitate either surface or underground drains. There is of course a public plaza—no Spanish-American town could be without one. A part of that of Sandy Point is shown in the photograph reproduced in Fig. 5, which also shows very well the ordinary and daily condition of the streets in even the better portions of the town. Although located in 53° south latitude, within five miles of a twelve-foot vein of good lignite and surrounded with dense forests, fire as a means of comfort is unknown. At the Cosmos Hotel, considered as the best in the town and where, for six dollars per day, I was permitted to occupy a small room with a single chair, bed and bare floor, and served with two very indifferent meals during the course of each twenty-four hours, there was no fire save in the kitchen. Upon my asking if any arrangements could be made whereby I might enjoy the comforts of a fire, I was told that, if cold, I could drink whiskey to keep warm. Acting upon this suggestion, I was thunderstruck to find that the charge for one whiskey and soda was eighty cents. It did not take long to convince me that, with the temperature as low as that of Sandy Point, I could not afford to burn such high-priced fuel. For my greater comfort, therefore, I took to my bed, which, considering my condition, was much the best thing I could do. After some two weeks passed with little of comfort, if not actual misery, I was sufficiently recovered to start on my return journey to rejoin Mr. Peterson. I reached the estancia of Mr. Halliday at North Gallegos late in November, where I found my companion, he having meantime finished the work near Coy Inlet. We immediately commenced making preparations for a protracted trip into the interior, selecting as our objective point the region of lakes Viedma, San Martin and Argentino. Mr. Peterson during my prolonged absence had been very successful, having secured nearly a ton of fossils. In our two previous shipments we had included nothing but fossils. We now decided to pack up all our remaining material, both fossil and recent, and leave it with Mr. Halliday for shipment to Sandy Point by the first vessel that should come in bound for that port. When all was made ready we had about six tons, which we left to be consigned to Braun & Blanchard, and forwarded by them to New York.

5—A Street Scene in Sandy Point.

6—Lava Beds at Palli Aike

FIGURES 5 AND 6—SEE OTHER SIDE

CHAPTER VI.

*Start on an extended trip into the interior; Select Lake Argentino as our
first objective point; Shoeing horses; Crossing the high pampa; A
splendid mirage; Scanty vegetation; Lava beds; Digging for water;
The Santa Cruz River; An attempt at fording the Santa Cruz; Rio
Bota; Lake Argentino; Crossing the Santa Cruz River.*

O N the thirteenth of December, having supplied ourselves with eight
months' provisions, we started on our trip into the interior. We
left our tent, stove, and such other articles as we were able to
dispense with, in order to lighten our load as much as possible. In lieu
of a tent we took with us two extra tarpaulins, from which we later constructed a tent. The first day we went only as far as Killik Aike, where
we stopped for the night with Mr. and Mrs. Felton. On the fourteenth
we went to Guer Aike, where I remained for the night, and Mr. Peterson
proceeded on horseback to Gallegos, to ascertain if a steamer which had
just come in from Buenos Aires had brought us any mail, for as yet we
were still without news from home. We had agreed before separating
that we should travel independently until meeting on the second day following at Governor Mayer's estancia on Coy River. At about noon of
the day agreed upon, December the sixteenth, I drove up to the estancia,
where I found Mr. Peterson awaiting me. After breakfasting with General
Mayer and Señor Villegrand, who chanced to be at the estancia, and bidding them good-bye, we resumed our journey. We little thought, as we
exchanged farewells with him who had ever extended to us and our work
a kindly interest, on that beautiful summer's day, that it was to be forever. Not only did our host seem in excellent health and spirits for a
man who had passed his sixtieth year, but in conversation that very day
he told me how he expected to live until past ninety by adhering to certain principles of living, which, he said, had been followed most sucess-
fully by the elder Burmeister, of whom he was a great admirer.

95

We had been told that two young Englishmen at Lake Argentino, were possessed of a boat by which we could cross the Santa Cruz River.   This information led us to choose this route rather than that by way of the mouth of the Santa Cruz.   This afterwards caused us no little trouble and delay.

For the first day and a half after leaving General Mayer's estancia we followed up the north fork of Coy River, or Rio Aubone of the map.   On the evening of the second day we came to a place where the stream made a long detour to the westward, and since our course lay in a northwesterly direction, we decided to leave the river valley at this point and strike out across the high pampa which lay beyond.   Since the grass was exceptionally good in the valley, we decided to stop for a day, in order to allow our horses to rest and recruit themselves for their long trip across the pampa. We had not supposed that we were nearer any human settlement than that of Governor Mayer's when, about ten o'clock on the morning of the following day, we saw a man on foot coming toward us from the opposite side of the river.   This was certainly unusual in a country where everybody disdains to walk for even the shortest distance, and it awakened in us no little interest and curiosity.   The stream was not deep and, on reaching it, he walked directly into and through it.   As he approached we made out by his gait that he was not an Indian, although as he came nearer, his clothing and hair were quite as unkempt as we had been accustomed to see among the Indians.   When he drew near I was quite unprepared to hear him address us in not only distinct but exceptionally good English.   He at once introduced himself, and explained that he lived just around a point on the opposite side of the valley and had been out looking after a pair of work-cattle, when he caught sight of our unfamiliar equipage and came over to see what manner of folk we were, for he had been long enough in Patagonia to acquire all the curiosity of the frontiersman.   After learning our intentions, he pointed out the best crossing of the creek, told us what he knew of the high pampa between us and Lake Argentino, which in truth was not much, and on taking his leave extended to us a very hearty invitation to move over to his place, where he assured us there was a spring of fine water, an abundance of fresh vegetables and welcome for so long a time as we chose to remain.

On the eighteenth of December we moved across the river to the house of our new acquaintance and stopped for the day.   With his wife (a half-breed Tehuelche woman) and one child, he lived in an adobe house of a single room. The roof was of rough boards with a hole left in one corner to permit the escape of the smoke from the open fire beneath, though in truth the greater part of it remained within, entirely filling the room. A small table of rough boards, one or two stools and a few tin vessels were the only articles of furniture of which this house could boast. It was not nearly so well furnished or comfortable as many of the toldas belonging to Indians of pure breed with which we had met while in Patagonia. I could not avoid wondering, as I sat and surveyed the poverty stricken and filthy nature of everything about me, what had brought this apparently well-bred man to such perfect contentment with his miserable surroundings. When I learned subsequently that he was not only from a good family, but had enjoyed the further advantages of a university training, my wonder was increased rather than diminished. Surely from a university life to this he was leading in Patagonia was a long step, but he appeared, except for his refined language, so completely in harmony with his present environment that I doubt not he had taken to them of his own volition and found in the half savage life he was leading a certain satisfaction, though little, I fancy, of comfort or pleasure. To me he furnished a striking example of the ease and rapidity with which representatives of the human race can relapse into barbarism from positions of high social and moral standing. Though morally, I am bound to say, that from all I could see or learn of him, there was little with which he could be reproached.

On the following day we moved on a few miles farther to some springs which gushed forth from beneath the bed of shingle on the side of the bluff some fifty or sixty feet below the surface of the pampa. In the night before, the mosquitoes had, for the first time during our trip, been very troublesome. At these springs we not only found most excellent water and an abundance of grass, but relief also from those pests which had been so annoying in the valley below. We passed the remainder of the day at this place, washing our soiled linen and shoeing our horses, preparatory to their long trip over the stony and lava-covered plain. Horseshoes were scarcely known in Patagonia. We had with great difficulty procured six of these very useful articles in Gallegos, along with a number of mule-shoes which had been shipped with other government stores from Buenos Aires to the commissariat. These had been knocking about among the stores for a number of years, not that they had never been needed, for almost daily one could see horses going about the streets lame from having their hoofs worn down too closely, but because of the indifference or want of skill on the part of those whose duty it was to look after such matters. However, this was extremely fortunate for us, as it put at our disposal without cost a supply of shoes, which we were able with little difficulty to fit and set as needed throughout our trip. This enabled us to travel steadily for days at a time over stony country without loss of time from maimed and sore-footed horses, as would certainly have happened, had we been compelled to undertake our journey with no means for keeping our horses shod.

Early on the morning of the twentieth of December we resumed our journey. We soon gained the top of the bluff, where we stopped for a moment to give our horses a breathing spell. Behind us the broad, deep valley of the river stretched away to the southeastward far as the eye could reach. The course of the little stream fringed with a narrow border of grass appeared as a delicate line of green pencilled on the dull brown surface of the surrounding landscape, as it meandered in graceful curves from one side to another of the valley. In front of us stretched the great plain ; its broad, level surface appeared as though interrupted by neither elevation or depression, as it swept away to the northwest, until finally blending with the distant horizon. The morning was such as would have been considered beautiful in any country ; for Patagonia it was exceptionally fine. The sun rose brilliantly out of the east into a deep blue sky. The atmosphere was clear and bracing, with a slight haze along the western horizon. The atmospheric and other physical conditions were just such as serve best to produce those wonderful mirages and other optical effects which are only seen at their best in semi-arid regions. As we continued our journey across the plains, the isolated calafate bushes and scattered patches of mata negra, some of which latter were of no small extent, became greatly magnified, and appeared as giant trees or outliers of a magnificent forest. A band of guanaco galloping across the plain with heads erect had the appearance of a troop of mounted cavalrymen, from which at a distance they were distinguishable only by their long, swinging gait, with which we had now become quite familiar. A slight elevation on the surface of the plain became magnified in such manner as to appear as an insurmountable barrier, so interposed as to interfere with our further progress. At first, until we discovered their true nature, these would cause us to diverge from our direct route. On drawing near, however, they would gradually disappear and in most instances a gentle incline of only a few feet would be found where we had supposed there was a precipitous bluff rising several hundred feet above the plain. At other times there would be double images of the same series of objects, one image appearing directly above another, owing to the different refracting powers due to the varying density of the several strata of the atmosphere which lay between us and the various objects. These interesting phenomena continued with most bewildering effects until about ten o'clock. Then they entirely disappeared and the bushes, animals, and inequalities upon the surface appeared in their normal conditions much to our relief, I may say. For while we had been intensely interested in it all, it was not without its annoying features, since it was very difficult to hold to a definite course, through the impossibility of being able to select any fixed landmarks, by which to steer a direct course. Nothing was stationary ; everything was constantly changing.

When, shortly after ten o'clock normal conditions were restored, we found ourselves in the midst of a vast plain with no permanent landmark visible in any direction. The horizon described a perfect circle, and within the circumference there appeared not even a solitary hill. The impression was not unlike that at sea in calm weather, as one scans the surrounding horizon, with this exception, that, at sea, I never have even the slightest idea as to the different directions until looking at the compass. Here on the plains of Patagonia, as on our own plains, I seemed instinctively to know the points of the compass, and could travel for days at a time without consulting that instrument.

In the afternoon the blue skies of the morning were overcast with dark, sombre clouds, which effectually shut out the genial warmth of the sun. The wind grew stronger and the temperature lower, so that with my position as teamster, I was chilled through and altogether decidedly uncomfortable when, late in the evening, we came upon two small water holes. As this was the first water we had seen during the entire day, we were not long in turning out and making ourselves ready for the night. There was a rather scanty supply of brown and withered, but nutritious grass about the pools, enough, however, to furnish sufficient feed for our horses for the night. As for ourselves, we soon had a brace of plover, well salted and peppered, and broiling before a bed of embers. These, with an abundance of good bread and a pot of hot coffee, were sufficient for our present needs. Supper finished, the contents of the cart were carefully covered for the night, and, after reassuring ourselves that the horses were all securely hobbled or picketed, we unrolled our beds alongside the cart, undressed, crept into the blankets and found warmth and rest, secure from the chilling blast that howled mournfully without.

The following morning broke cold and uninviting. The atmosphere was raw and damp, and continued so throughout the day. We were on our way early. Hour after hour the cart rolled wearily over the shingle-covered surface. The vegetation was extremely scanty, and such plants as were to be seen belonged for the most part to those fleshy, leafless varieties common to arid districts. Besides such, there were a few inconspicuous members of the Compositæ and Leguminosæ, while a member of the Cruciferæ with rather large yellow flowers seemed to thrive fairly well, since I saw frequent examples of a foot or more in height. Large circular masses of *Bolax glebaria* were everywhere.

Early in the forenoon we saw, through the clouds outlined on the horizon in front of us, a number of small conical elevations. As we drew nearer, these proved to be small outlying craters, or lava vents. Each of these was surrounded by a lava bed of limited area, similar to that shown in Fig. 6. For several miles we wound in and out among these local lava sheets, finally gaining the level pampa which lay beyond. Late in the afternoon we came to a deep draw, or cañon, with rather precipitous but grass-covered slopes. Since we had seen no sign of water throughout the day and the country ahead was equally unpromising, we decided to follow the course of this cañon for a distance, in hopes of finding sufficient water for the night. Just as it was growing dark, we came upon a place where the nature of the grass at the bottom of the cañon indicated that water was present on the surface, or could be had by digging, at no great distance beneath. Driving the cart to a narrow bench just beneath the crest of the pampa, in order to secure shelter from the wind, we turned out for the night. Although there was no water at the surface, we obtained sufficient for our needs, by digging, at a depth of about three feet.

I had felt miserable throughout the day. The wound on my head, which had not yet entirely healed, was paining me considerably, and since I had no appetite for supper, immediately that the horses were cared for, I sought the comfort of my bed, where I passed a restless night. On the following morning, December twenty-second, I had developed a high fever and my head about the old wound was swollen, very much inflamed and quite painful. I was clearly unfit to go on and decided to lie over for the day until the fever should subside. We had no remedies, so that I was quite without medical assistance, except for a few five-grain phenacetine tablets. I began and continued taking these until they were exhausted, but, despite every effort and the kindly care of Mr. Peterson, the fever continued to increase, while the wound on my head reopened. The swelling spread so rapidly over my head, face and neck that on Christmas day both my eyes were entirely closed. Mr. Peterson repeatedly urged me to allow him to make a couch in the cart and return with me to Gallegos, but I could not think of this. Until the time of my injury while on the road to Sandy Point I had seldom known what it was to be ill, and I did not now for a moment doubt my ability to pull through. After several days passed in a half delirious state and suffering from fever and the erysipelas in my head and neck, I began to mend. The inflammation subsided rapidly, taking with it, however, most of my hair, and on the first of January we were able to resume our journey. During those two or three days of my sickness, when I was suffering most with the terrible fever from which I could get no relief, passing hours at a time in a half delirious state, I frequently had most vivid but imaginary experiences. One of these I wish to relate in this connection, because of the permanency of the impression which it produced upon my mind. I fancied myself as going on an expedition to Greenland, a place, by the way, where I had never had the slightest desire to go. The personnel of the party, the name of the ship and her officers, as well as our landing place and the site of our encampment in Greenland were all, not only brought distinctly and vividly before my imagination, but so deeply and thoroughly fixed were they in my mind, that it was many months after my recovery before I was quite able to convince myself that I had really never been to Greenland.

During my convalescence Mr. Peterson had passed some time in reconnoitering. He discovered that the Santa Cruz River lay only a few miles to the north of us, and that consequently we had directed our course across the pampa rather too much to the north and not quite enough to the west.

Leaving that bleak and desolate spot upon the pampa, where I had all unwillingly passed the most miserable Christmas of my life, with the new year of 1897 before us, we set out over the plain, holding our course a little north of west.   We travelled all day across the pampa, passing a number of small lava beds similar to those described above, and camping at night by the side of a small lake, which occupied a shallow depression in the surface of the pampa only a few rods from the crest of the bluff, which from an altitude of twenty-five hundred feet overlooks the valley of the Santa Cruz.   Owing to the gentle southeasterly slope of the plain there had been no indication of the valley, until we arrived at the brink of the precipice and the whole panorama lay spread out before us.   To the westward the valley expanded into a great basin, at the farther end of which the blue waters of Lake Argentino were visible, its western shores extending far back among the rugged, snow-capped peaks of the Cordilleras. To the eastward the great valley extended as far as the eye could reach, like a giant furrow, ten miles wide and half a mile deep, plowed in the surface of the plain.   The bottom of the valley was as brown as the plain above.   Not a patch of green was there anywhere to relieve the monotony of the color.   From our point of view the mighty river appeared like a silver thread, as it wound back and forth from one side to another of the valley.   The great plain over which we had been travelling stretched away to the south, as if limitless in its expanse, while north of the river lay the boundless plains and lava beds with their hundreds of picturesque volcanic cones, in the midst of which we hoped soon to be.

As the sun rose bright on the morning of January the second, the view from our camp on the crest of the bluff, overlooking the broad valley of the river and Lake Argentino, was most beautiful.   The lake shone like polished silver in the bright morning sun-light.   In the distance the snow-capped peaks and ranges of the Andes, as though chiseled from purest alabaster, gleamed with dazzling whiteness, as each towering mass rose heavenward and lay silhouetted against a black background of sombre clouds, that hung threateningly over the Pacific slopes.

The precipitous nature of the bluff on which we were encamped made it necessary to reconnoitre a little, in order to find a place where we could safely descend to the valley with our cart.   Saddling a horse, I started out for the double purpose of finding a practical route to the valley below and of learning what I could of the nature of the valley and river and the surrounding bluffs.   Travelling along the crest of the incline, I soon came to a point where a small tributary entered the valley from the plain above, and the ridge or hogback separating this from the main valley proved a feasible, though far from excellent, highway for our cart. The slopes were so much covered over with secondary materials, that I could determine little concerning the nature of the rocks which lay beneath. The valley at this point had a width of perhaps ten miles, and as I rode across it to the river, I noticed that the surface, instead of being level, was covered over with numerous small rounded hillocks, each rising to a height of from twenty to forty feet and from one hundred to two hundred feet in diameter. On the summits, about the slopes and over the depressions between these hillocks, were numerous large angular blocks of gneiss, granite and other rocks, for the most part crystalline. I noticed that these hillocks were arranged in series, crossing the valley at right angles. Each series was separated from the one preceding or following by a narrow, level valley, running at right angles to that of the river. It was clear that the succeeding ranges of low hillocks were terminal moraines left by a great glacier that had formerly descended the valley, at least to this point, completely filling the basin now occupied by the great lake above, and the river valley as well for some miles below. Everywhere as I rode along through the hills and across the valley, I noticed that the surface was as dry and parched as that of the plain above, even more so, in fact. After a time I arrived on the bank of the river, at a point where, by a number of small islets, it is broken up into several channels. I had been told by Mr. William Clarke, a North American who had lived in the country for the last forty years, that somewhere near the source the Indians were said to know of a place where the river was capable of being forded. It occurred to me that if there was any such place anywhere along the river, this must be the place. It would be a very great saving to us if we could ford the stream at this point and proceed on our way north. Being myself no novice at swimming, I resolved to determine for my own satisfaction whether or not it was possible to ford the stream at this place. From the nature of the current I could easily detect the shoal waters, and by keeping a little on the upper side of the riffles I successfully crossed the first channel and with somewhat greater difficulty gained the second of the series of small islets that lay between me and the opposite shore. The third channel was considerably broader than either of the two I had placed behind me, and its waters rolled along with such an even and undisturbed surface that I could with difficulty detect any point of vantage from which to attempt a passage. At the upper end of the island there was, however, a faint ripple on the surface of the water, indicative perhaps of a riffle beneath. I chose this point, realizing the certain advantage it would afford for making a landing somewhere on the island, in case my horse should suddenly encounter swimming water, and thus avoid the possibility of being swept by the rapid current into the middle of the stream below the islands. Slowly and cautiously I urged my sturdy little horse into the icy water, which gradually gained in depth, until, in midstream, with the waters surging by my saddle skirts, I stopped for a moment to consider whether I had not best turn back, for, after all, if I succeeded in gaining the opposite shore on horseback, was it possible to cross our loaded cart through a stream so deep and swift? For an instant only I hesitated to essay the stronger current that rushed madly by in my front, then, holding my horse well up stream, I urged him forward. He had evidently stopped immediately on the crest of a high bank, for with the first step forward he plunged head downward into swimming water, and we were suddenly in the midst of the current, by which we were being swept rapidly down stream. The little animal made a gallant struggle, while I lent such assistance as was possible by throwing myself on one side and steering him for the island, in hopes that he might find bottom before we had passed its lower limits. In this hope I was fortunately not disappointed, for after a few brief moments which seemed all too long however, passed struggling in the icy waters, his front feet struck on the bar, and with an extraordinary effort he succeeded in gaining its surface. I was not long in regaining the island and recrossing the two channels I had successfully passed. Once on shore I realized the fact that, since I had only two days before left my bed, I could hardly be considered as beyond the stage of convalescence, so, drenched to the skin, I started on my return to camp at the summit of the bluff, fifteen miles distant. From this and several subsequent experiences with this stream I am content to abide by the description of it given by Darwin. On April thirteenth, 1834, he described it as a river from three to four hundred yards broad, with a depth in the middle of about seventeen feet and a current of from four to six knots an hour. It had apparently not deteriorated when I first visited it on January the second, 1897. It is a magnificent river, as it rushes swiftly over its stony bed from its source in Lake Argentino, at the base of the Andes, almost due east to the Atlantic, one hundred and seventy miles away.   Throughout its course it does not receive a single tributary of any importance.

As I galloped along rapidly on my return to camp, I noted the great constriction in the valley a few miles below and recalled Darwin's description of the valley of this same river at the highest point gained by Captain FitzRoy, when on the morning of May the fifth, 1834, he turned back after tracking three large whale boats for seventeen days up the river.   From the topography of the surrounding country and Darwin's description it was evident that their party had stopped at the bend in the river a little below and directly opposite our camp.   Describing the valley from their uppermost landing on the evening of May the fourth, 1834, he says, "The valley in this upper part expanded into a wide basin, bounded on the north and south by the basaltic platforms and fronted by the long range of the snow-clad Cordillera."   Although not more than twenty miles away lay the magnificent Lake Argentino, so near in fact that it would have been easily visible to Darwin and his companions had they ascended any of the surrounding elevations, yet they entirely missed it, and this pearl of the Andes remained for many years undiscovered, along with Lakes Maravillo, Viedma, San Martin, Nansen, Pueyrredon and a host of others, that combine to form a chain of mountain lakes that for size and beauty are unsurpassed, if not unrivalled, elsewhere in the world, and of which I shall speak more particularly when I come to treat of the lake-systems of Patagonia.

I had been delayed longer in the valley than I had thought, so that it was late in the afternoon when I returned to camp and changed my thoroughly drenched and cold garments for others dry and warm.   Fortunately I experienced no ill effects from my unwilling bath in the river, and the following morning, January the third, we descended with the cart and our belongings to the valley by the route I had selected, and encamped near the mouth of the Rio Bota, a small stream emptying into the river some twelve miles below the lake.   We found this stream abundantly stocked with fish ranging in size from a half to one and a half pounds.   As these fish were of excellent flavor and took the hook readily, they afforded us a most desirable addition to our menu. This genus, *Galaxias*, we afterward found to be common in all the streams of Patagonia.   It is about the size and shape of the brook-trout, of a light brown color, and with scales similar to but somewhat smaller than our lake white-fish. The quality of the flesh is not inferior to that of the speckled brook-trout.

On the following morning, January the fourth, I set out on horseback to find the house of the two Englishmen who, we had been told, lived at the east end of the lake. I was especially desirous of ascertaining what arrangements we could make with them for being conveyed across the river in the boat they were represented as having. I found no evidences whatever of any habitation at the eastern end of the lake, but following along the north shore in a westerly direction for about twenty-five miles I came upon the house in the edge of the mountains. Only one of the men was at home and he was suffering with a broken ankle and was attended by a couple of Chileno laborers. I was immediately made welcome to such meagre accommodations as the place afforded. When I asked concerning the boat, what was my disappointment on being informed that it had been abandoned two years previously at a point some sixty-five miles below the lake. It had been seen within the last six months by one of the men and its owner had no doubt that it still remained where they had left it. He assured me that if it could be of any service to us, we were welcome to it. I remained at this place over night and returned the next day, January the fifth, to camp in time to make the necessary arrangements for starting on the following morning down the river in quest of the abandoned boat, which we hoped might still be sufficiently serviceable to enable us to cross the river in safety. Before leaving our camp on the Rio Bota, however, we did not neglect to catch and salt down a supply of fish sufficient to last us for several days.

On the morning of January sixth we left the Rio Bota and, knowing it would be impossible to travel for any great distance along the river valley, we at once climbed the long and precipitous incline by the same route by which we had descended from the plain into the valley. We were all day gaining the summit, and a severe and trying task it was for our horses. We camped at night by a small lake at the top of the pampa and the following morning set out on our journey down the river.

Throughout the seventh and eighth of January we travelled steadily over the pampa, keeping for the most part near the crest of the bluff which overlooked the deep valley of the noble river. The surface of the valley, like that of the pampa, was everywhere parched and barren. As Darwin has truthfully said, "the curse of sterility was everywhere." While travelling down the river we passed at a distance several herds of wild cattle grazing on the dry, coarse herbage of the pampa. On the afternoon of January eighth we came to a place which in every way answered the description of that which had been given me of the place where the boat had been abandoned. Mr. Peterson rode down on horseback to make sure that it was the proper place and that the boat was still there, before we should venture down with the cart. He found the boat at the place we had suspected it would be found and returned to help me down the steep and perilous incline with the cart. May the good Lord forever preserve me from ever again being compelled to resort to a two-wheeled vehicle to get about over a rough and broken country. How I wished that afternoon and many times afterwards for a light mountain wagon fitted with a California brake. All the remaining afternoon was spent in the descent of the bluff. Owing to the gentle nature of our shaft-horse we were fortunate in meeting with no greater mishap than the breaking of the hold-back straps, and late at night we arrived on the bank of the stream near where the boat had lain for the past two years high and dry on a little sand bar at the mouth of a small draw that entered the river at this point.

On the morning of January ninth we were up early and busily inspecting the good ship that was to carry us to the opposite shore. We noted carefully her many defects and discussed thoroughly the best means for remedying each. We were not long in agreeing that, while not entirely seaworthy, she could be made to answer our purpose. Covered by an old sail there remained in her a considerable quantity of sugar and perhaps two dozen bottles of Worcestershire sauce. The presence of the sugar was proof positive that she had suffered chiefly from drouth, as her gaping sides only too plainly indicated. In and about her were a small mast, a number of oars, and several boards. By the aid of the oars and mast as levers we succeeded in turning her keel up, when we calked and battened the cracks as best we could. Then, righting her again, we applied such additional remedies as was possible from the inside, and with some difficulty slid the old hulk down into the water, using the oars as ways. As we had expected, she instantly filled. Making her fast to the shore we left her in this condition to soak until the morning of January thirteenth. We passed the intervening time in examining the strata of the bluffs about us, (which we found to consist of materials belonging to the Santa Cruzian formation), in constructing a small tent from our tarpaulins and poles from the bits of lumber, and in an excursion to the pampa above in quest of wild cattle from which we secured a bountiful supply of most excellent beef. On the morning of the thirteenth we baled the boat out and found that, though still leaking badly, she could be made to float long enough to cross to the other side. Having previously selected an advantageous place for loading, we had taken our cart to pieces and stored it along with our other effects on the shore, conveniently arranged in the order we had decided that each should be placed in the boat, so as to insure against damage such articles as were likely to receive injury from getting wet. We next hauled the boat up alongside and made her fast. Our horses were then brought down and driven into the river. They took to the water with some reluctance, but by yelling at them with all our might and pelting them with stones, we at last succeeded in getting them headed for the opposite shore, where they finally landed, though at a considerable distance below. It was now our turn. With our motive power on one side of a stream three hundred yards wide and seventeen feet deep, and ourselves and equipment on the other, something had to be done. We again baled out our boat, hastily embarked ourselves, cast loose and pushed her off into the current. With Mr. Peterson pulling with all his might at the oar on the lower side and myself at the rudder we succeeded in keeping her headed up stream and for the opposite shore, until she struck and grounded on a bar distant some two or three rods from the north bank. All our efforts to get her off were futile, and we were compelled to carry our belongings to shore through the water rising about us to our waists. However, we were now safely across the most considerable river of southern Patagonia and with little or no loss to our equipment. We remained on the bank of the river for the rest of the day, setting up our cart, drying such articles as had become wet in crossing and making such other preparations as were necessary for our start on the following morning.

CHAPTER VII.

*Through the lava fields; Lavas not of submarine origin; Rio Chalia or
Sheuen; Bad lands in Santa Cruz formation; The finding of a human
skeleton; In the valley of Rio Chalia; Mosquitoes; The Armadillo;
Sierra Ventana; The Rio Chico and its basalt-capped cañon; Cyanotis
rubrigaster; The Patagonian mocking bird; Burrowing owls; Owls and
rodents; Protective coloration in sand lizards; Character of river cañon
and basaltic platforms; Barren nature of the lava beds; Our first view
of the Andes from the valley of the Rio Chico; A terminal moraine;
Abundance of rodents; Cavia australis; Ctenomys magellanica, the
Tuco-tuco; Numerous mice; Terrific rainstorm; Destruction to animal
life wrought by storm; Effects of burrowing animals on erosion; Ascend
south fork of the Rio Chico; Glacial moraines and lakes along the upper
stretches of the river; Enter Mayer Basin by way of Shell Gap; On
the outskirts of the Andean forests.*

THE country to the north of the river at this point had the appearance of being much more broken and obstructed by lava beds
than that to the south. So that we anticipated some difficulty in
getting up out of the river valley upon the pampa. Before starting, on the
morning of the fourteenth I rode out on horseback to inspect the bluff and
find, if possible, a practical route out of our present situation. I was not
long in discovering such, and soon returned to camp, when we immediately set out on our journey northward. We had soon crossed the barren,
shingle-covered valley of the river and were ascending the steep, rugged,
lava-covered slope to the north. By winding in and out among the projecting ledges and taking advantage of every incline that offered a promising trail to the summit of the next higher platform, we gained the top of
a low table which lay between the river valley and a narrow, deep basaltic
cañon, which entered the latter a little way below. Keeping along on the
surface of this table for a distance, we came to a point where the bottom
of the cañon opened out into a broad open stretch of meadow-land, per-

haps a quarter of a mile in width.   On the opposite side there appeared
a steep but gradual slope from the bottom of the cañon to the high pampa,
where we should be out of the limits of the basalt.   In front was a practical route by which we could descend to the bed of the cañon, so that in
a few moments we found ourselves at the bottom, which opened out
above into such a splendid meadow of beautiful waving grass, fresh and
green, with an abundance of pure sweet water, that we could not resist
the temptation to turn out and pass the remainder of the day and night
here, where our horses could be on such excellent feed and water and find
time to recuperate after their hard pull up the steep, rocky slope of the
river bluff strewn with broken fragments of basalt.   Moreover I wished to
spend a few hours in observing the basalts and studying their relations
to the underlying sedimentary rocks.   I had noticed during the morning
that between the valley of the river and the high pampa which lay some
miles to the northward there was a series of terraces, and that each of
these down to a certain level was capped with basalt and appeared like
successive basaltic platforms.   I was interested to know whether these
represented successive flows at different intervals, or one flow which had
taken place at a comparatively late period after the river and its tributaries
had carved their valleys down at least to the level of the lowermost of
the lava-capped benches.   I discovered that in this locality at least there
had been but the one flow, and at different points along the sides of the
small cañon where sufficient erosion had been effected at opportune places,
the different layers of lava could be seen as they flowed over the crest
and down the slopes of the various inclines.   By following up the cañon
for the distance of a mile the little park in which we were camped was
seen to terminate suddenly.   The walls contracted on either side until
the park was reduced to a deep, narrow cañon only a few yards in width.
At the summit of this extremely narrow gorge there were two narrow,
basalt-covered platforms extending back to the slopes of the broader main
cañon above.   It was very evident that the position of the basaltic layer
at the top of the lowermost narrow gorge indicated the extent to which
erosion had taken place at the time of the outflow, while the gorge
beneath had been entirely cut subsequent to that time.   On ascending to
the summit of the high lava-capped plain to the westward, I found the
surface deeply riven, presenting an extremely rough and broken surface, with deep crevices and great yawning caverns.   Nowhere, however,

at this point did I see any shingle upon the surface of the lava. It was, therefore, clear that this particular stream of lava had been ejected subsequent to the deposition of the great shingle formation of Patagonia. The fact that no shingle or other detritus is found upon the surface of the lava beds throughout most of Patagonia appears to me as conclusive evidence that these lavas were not poured out over the bottom of the sea as supposed by Darwin. All these matters will be discussed later when I come to treat of the geology of Patagonia.

On the fifteenth of January we left our camp at the bottom of the cañon and, pulling slowly up the eastern slope, finally gained the summit. To the north of us lay a series of low, terraced benches, or steps, each rising successively higher, until at a distance of some twenty or twenty-five miles the summit of the high, level pampa appeared. From our point of vantage on the east side it was evident that we had acted wisely in crossing the cañon at this point, for not only would it have been quite impossible to have effected a crossing for many miles above, but the nature of the surface on the opposite side was such as would have offered insurmountable obstacles to our further progress. Far away to the north and west, on the western side of the gorge, there extended a continuous and uninterrupted lava sheet. As far as the eye could reach there appeared only the broken and uneven surface of black and barren rocks. The surface of the plain at the higher levels on the east side of the cañon was free from lava and afforded a fairly practicable highway to the northern country, for which we were headed. The topography of the country and the distribution of the lava bore positive evidence that the greater part of the cañon had been cut prior to the outflow which had submerged the plain to the west. A number of old craters seen in the distance had undoubtedly been the source of the lava. From these it had flowed down in great streams over the gently inclined surface, until coming in contact with the eastern wall of the cañon which had confined and restricted the flow in that direction.

By keeping along near the eastern crest of the cañon wall we found fairly good travelling, and ascended the successive terraces as they were encountered. Although constantly increasing our altitude, the bottom of the cañon below was rising even more rapidly, so that at a distance of twenty miles, or just before gaining the broad level surface of the higher pampa, the previously rugged nature of the cañon disappeared and it expanded into a broad, level basin-shaped valley with numerous fine springs of water and a bountiful supply of grass.   By the side of one of these springs we camped for the night, putting our horses to graze in the valley beneath.   A fair conception of the nature of the basaltic cañons of Patagonia can be had by inspecting the accompanying illustrations in Figs. 7, 8, 9, 10, made from photographs taken by the author. Walking about this little valley during the evening and examining the various pools and springs to ascertain what they contained of animal life, I was struck by the great abundance and character of a species of Chara with which every water hole and spring was literally filled. In collecting this plant for the herbarium I noticed that it was extremely harsh and brittle, even while still wet and fresh from the pool.   So highly charged was it with mineral matter ($CaCO_3$) that it was difficult to separate and properly arrange the different specimens in my drying papers without their breaking up into little bits.   On using the water it was found to be very hard.   The water of these springs was so highly charged with carbonate of lime, that all the pebbles and rocks lying about the water's edge were coated over and often cemented together with a thick covering of that mineral.   Following the course of several of these springs to the bed of shingle and the basaltic platform above, I found the basalt full of cavities filled with crystals of calcite.   This explains how the water of the springs beneath had become so highly charged with carbonate of lime.   Over this little valley and about the foot of the basaltic cliffs, which for the most part surrounded it, there grew a number of plants which I had not observed elsewhere, and the remainder of the evening was passed in securing and pressing a set of these as additions to our botanical collections.

Early on the morning of the sixteenth we resumed our journey.ˑ We were not long in climbing the slope which lay between us and the broad, high pampa above.   When we had gained the surface of the latter, we laid our course in as nearly as possible a due northwesterly direction. Travelling for a distance of from twenty-five to thirty miles across this pampa, the surface of which sloped gently from northwest to southeast and was entirely unrelieved by either elevation or depression, at about one o'clock in the afternoon we came suddenly and unexpectedly to the crest of a high escarpment overlooking the valley of the Rio Sheuen, or Chalia, which lay some two thousand feet below.   Immediately in front

7—Cañon of Arroyo Gio.

8— Cavern in Lava Beds.

9 – Basaltic Pinnacles

10 – Fissures in Basalt

of us was a considerable "bad land" area, not unlike, in general appearance, those of the Oligocene deposits of Nebraska and Dakota, and quite similar to that shown in Fig. 12. It was of great importance that we should explore these, at least sufficiently to determine what fossils they contained and to what formation they belonged. Reconnoitering for a moment, we discovered a large hole of water in a basin at the top of a landslide a few hundred feet below us. Since this seemed as practicable a route as any by which to descend to the valley, with some little difficulty we succeeded in getting our miserable two-wheeled cart down the steep incline to the water, where, after caring for our horses and refreshing ourselves with a hearty meal, we set out to spend the remainder of the afternoon in exploring the bad land area mentioned above. Hardly had we reached the first exposure when I picked up a beautiful little jaw, with black, shining teeth, of *Abderites crassiramus*, a small extinct herbivorous marsupial. Notwithstanding this fortunate discovery, the locality proved not especially favorable for collecting vertebrate fossils. We obtained enough, however, to show that the beds for the most part, at least, belonged to the Santa Cruzian formation, from which we had already made such extensive collections along the coast at Killik Aike and between Cape Fairweather and Coy Inlet. During the course of my wanderings throughout the afternoon, in the bottom of a deep and exceedingly narrow bad-land gulch, I came upon the remains of a human skeleton. From the condition of the teeth and the epiphyses of the limb bones I should judge it to have belonged to an individual perhaps nineteen or twenty years of age. The weathered and decomposed condition of the bones bore unmistakable evidence as to the great length of time they had lain in their present position, while the ghastly hole in the side of the cranium spoke only too plainly of the terrible tragedy which had taken place in this lonely and unfrequented spot. The skull was carefully removed and is now among the ethnologic materials in the United States National Museum at Washington. The bones of the trunk and limbs were in such an advanced state of disintegration as to render their preservation difficult, if not impossible. They were, therefore, left to complete the process of decay in their original resting place.

On January seventeenth we descended the slope to the valley of the river Sheuen or Chalia, stopping for a few hours at the base of the bluff to examine some marine deposits, which we had observed underlying the Santa

Cruz beds at this point.   In the afternoon we attempted to cross to the north side of the river and valley, but owing to the swampy nature of both, found this impossible.   We then directed our course up the valley, in order to find a more advantageous place at which to cross to the other side.   We soon discovered that the impediment offered to our progress by the usual swampy condition of the valley throughout this part of its course had been greatly increased by recent heavy rains, which had left the entire surface covered with mud to a depth of several inches, into which our horses and cart sank to a depth that made our further progress exceedingly tedious and fatiguing.   We were forced to pass the night in the valley alongside the marsh.   Throughout the night our horses and ourselves were almost literally devoured by mosquitoes, which swarmed in millions from the surrounding swamps.   All night long and until late in the morning of the following day they continued unrelenting, making night hideous with their chorus composed of a multitude of notes.   Try as we might, we could not keep them from getting in under our tarpaulins and bed-clothing, where they unremittingly engaged in inflicting upon us such wounds as only these insects can.   They were so troublesome that early in the evening it became necessary for us to securely picket our horses, lest they should stampede and become lost during the night.   All night long while almost smothering from having our tarpaulins and blankets tucked about our heads in such manner as to afford all the protection possible, we could hear our poor horses rolling and moaning outside, driven to distraction by the myriads of pests.   As daylight broke on the morning of the eighteenth, we left our beds and, without stopping to prepare breakfast, started on our way up the valley hoping to get beyond the limits of the swamp, where we might reasonably expect to find relief from the continual annoyance of our intolerable tormentors.   The morning was the most sultry I remember to have experienced in Patagonia. The clouds hung low and thick, without a breath of air that could in any way be likened to a breeze.   All the morning, as our faithful animals dragged the heavy cart slowly through the mud and water of the valley, these insects pursued their nefarious operations.   At about ten o'clock the clouds lifted a little, and a most welcome breeze sprang up from the west.   This rapidly increased in force, so that almost instantly the mosquitoes disappeared entirely and we were left to pursue our journey unmolested.   At midday we had passed the uppermost limits of the marsh and, finding a place where the stream, with low banks on either side, flowed over a bed of coarse shingle in such manner as to offer an excellent natural ford, we crossed safely to the other side, where, with an abundance of grass and pure water from the stream, we camped for the remainder of the day and night, in order to gain for ourselves and horses that sleep and rest we so much needed after the experiences of the preceding night. During the afternoon Mr. Peterson shot and skinned a few birds, while I, with my plant press, proceeded to a rocky ledge that projected out into the valley from the east at a distance of a few miles below. I found this to consist of a dike of hard basaltic rock protruding from the surface at right angles to the nearly horizontal strata of the surrounding sedimentary deposits.

Just above where we were encamped for the night the valley of the river curved sharply to the westward and opened out into a broad basin. On the western and northwestern border of this we could see extensive exposures which, from their colors, evidently belonged to formations other than any of those we had previously examined. On the following morning, January the nineteenth, we moved some fifteen miles farther up the river, and spent the afternoon in an examination of the exposures mentioned above. We found these to consist of several hundred feet of rather coarse, brown and variegated sandstones barren of fossils excepting a few uncharacteristic plant impressions. The following morning, January the twentieth, was passed in studying some lower hills, a few miles to the east. These proved to be made up, for the most part, at least, of marine deposits belonging to the great Patagonian formation. Returning to camp, we prepared an early dinner and left the Rio Chalia that same day, travelling almost due north across a series of low pampas and over the beds of several old dried-up lakes, that occupy depressions in the surface of the former. Our course lay along the foot of a high escarpment capped by a basaltic platform that rose to a height of from six to eight hundred feet above the plain to the east, upon which we were traveling. At night we camped at a small spring near the mouth of a small cañon that emerged from the high, basalt-capped bluff on our left and entered a deep dried-up lake basin that lay on our right. The spring afforded water just sufficient to supply us and our horses. Having noticed some promising exposures about the shores of the desiccated lake, I was up early the next morning, and, on examining them, found a number of marine invertebrates that I had not previously noted elsewhere. After an hour or two spent in collecting from these exposures, we resumed our journey, travelling northward all day and surrounded by a landscape with the same general features as that of the previous day. The high basalt-capped cliff on our left would extend due north in places for several miles and then sweep around far to the west, almost enclosing a great bay or gulf, so to speak, across the mouth of which our course would lead us. The low plain across which we were traveling supported a much better growth of grass than the high pampas to the south of the Rivers Chalia and Santa Cruz. The animal life was likewise more abundant. Bands of guanaco numbering from sixty to two hundred were common, while the *Rhea darwini*, either singly or in flocks of from ten to twelve, were scattered about everywhere over the plain. Perched in conspicuous places along the lofty basalt cliffs were numerous condors, while carranchas were, of course, not wanting. In addition to all these and many other birds and mammals, which had been our daily companions ever since our arrival in Patagonia, there was one curious little mammal belonging to an entirely different order, representatives of which we had not met with south of the Santa Cruz River. I refer to the little armadillo, *Tatusia hybrida*. Frequent examples of these were to be seen running about over the pampa or lying prone upon the ground. Immediately on touching one of these little animals, they roll themselves up into a compact ball in much the same manner as do some of the leeches or species of chitons, on being detached from the stones to the surface of which they are usually fixed. When in this position, the bony covering of the carapace serves to protect them from their ordinary enemies. They live in shallow holes excavated in the surface of the pampa, and if by any chance they succeed in reaching the mouth of one of these before being captured, they force the serrated edges of the carapace into the surrounding dirt in such a manner that they can be extracted only with the greatest difficulty. At this latitude they hibernate in winter and prefer a warm sandy soil and sheltered locality. In such places they are fairly abundant north of the Santa Cruz River, but we never observed a specimen south of that stream, nor after careful enquiries could I discover that they had ever been seen by others in the region lying south of this river. It seems probable, therefore, that this stream has afforded an effective barrier to their further distribution to the southward, for not only are there many localities to the south that would

Figures 11 and 12—See other side

11 — Sierra Ventana

12 — Bad Lands near Mt. Observation.

seem quite as well adapted to their needs as those to the north, but the entire southern half of the valley of that river is especially well suited to them. Though common in the valley on the north side of the river, no example has ever been taken to my knowledge in the valley on the south side. The temperature of the water in this stream, its great size, and the absolutely treeless nature of the entire region through which it flows render it particularly capable of presenting an effective barrier to the free migration of certain mammals, and more especially those like *Tatusia*, which are probably not capable of swimming and are known to hibernate in winter, at which period alone they would be able to cross such a stream on the ice. Their flesh is of an excellent flavor and highly prized by the natives as food.

On the evening of the twenty-first of January we camped at a small spring on the western border of a bad-land area of considerable extent. An hour or two spent in examining the various exposures in the neighborhood of our camp resulted in the discovery of sufficient material to fix the deposits as belonging to the Santa Cruzian formation.

On the twenty-second we crossed through these bad-lands, stopping at various places to examine such localities as appeared especially promising for fossils. In the evening we camped at a nice spring on the south side of the River Chico and some three miles above Sierra Ventana, a photograph of which is shown in Fig. 11. This mountain, which is in fact the core of an ancient volcano, stands directly on the south bank of the river, above the valley of which it rises to a height of some twelve hundred feet. During the afternoon we ascended this gigantic mass of basalt and cinders and stood on what still remained of the rim of the crater which crowns its summit. From the bed of the river the first five hundred feet or more of the slope consists of sedimentary materials belonging for the most part to the Santa Cruzian formation. The surface of this lower part of the slope is almost entirely covered over with great blocks of columnar basalt, and quantities of fine cellular pumice that have fallen down from above. At an altitude of some six hundred feet from the base the giant columns of basalt, which now fill the former vent of the volcano, rise out of the surrounding sedimentary rocks to an additional height of some four hundred feet. These columns gradually converge toward their summits, as shown in the photograph reproduced in Fig. 11, as though there had been a constriction in the neck of the volcano just beneath the crater. Resting upon this columnar mass of basalt are some two hundred feet of heterogeneous igneous materials consisting of great masses or lenses of obsidian and heavier slags near the base, and passing into highly vesicular pumice and scoriæ toward the summit, the whole exhibiting a striking variety of most brilliant colors, jet black, steel blue, vermilion and ochre predominating. From the summit we obtained an excellent view of the surrounding country. To the south in the foreground lay the low, level pampa which separates the Rio Chico from its smaller tributary the Rio Chalia, while beyond the latter stream rose the high pampa over which we had travelled on our way north from the Santa Cruz River. To the north and west of us, and extending along the north side of the river far to the east, were lofty and rugged basalt-capped table-lands. Of the exact nature and extent of these we had at that time no adequate conception. The valley of the river stretched away to the east, bounded on the south by a low, grass-covered plain and on the north by the basalt-covered table just mentioned, which rose to a height of four-teen hundred feet. To the west the river could be seen emerging, at a distance of about ten miles, from a deep cañon with walls on either side capped by lofty, precipitous basaltic platforms, towering fourteen to fifteen hundred feet above the stream. Within a radius of a few miles from where we sat there were no less than a half dozen old volcanoes all similar to, and equalling in beauty and interest, the one we had climbed. As I sat and surveyed the surrounding scene from my point of vantage, I was deeply interested in all the various phenomena about me. Each showed on what an enormous scale nature engages in her work both of construction and destruction. I could not resist the temptation to reconstruct in my mind the appearance of the surrounding country during the different epochs through which it had passed in arriving at its present, to say the least, half desolate condition. How different from the present were the condi-tions when, during the process of the last elevation of this region above the sea, the low plain which now lay to the south was still submerged beneath a great bay from the Atlantic, in the midst of which the pin-nacle upon which I sat was a small, rocky, wave-swept islet. Then again, how different from either was that other, and as it appeared to me, more remote period, when the volcano on the lip of whose crater I now sat in security, together with the others in the surrounding neigh-borhood, were in a state of violent activity and the great masses of igneous materials, which now cover to a depth of many feet the surrounding country, were ejected from numerous fissures and poured out over the surface of the plain!   How far had these phenomena been responsible for the extermination of that rich and varied fauna that had lived and flourished in such abundance in Santa Cruzian times?   While engaged in collecting their remains, I had noted that for the most part, judging from the anatomical characters exhibited by their bones and teeth, they had been well adapted for sustaining themselves in the midst even of only fairly favorable environments.   What then had caused the wholesale destruction or disappearance, not only of genera and species, but of entire families, and, in some cases, even orders of these animals?   Had their extermination been gradual, or had they been overtaken and completely destroyed by some sudden catastrophe?   If the latter, was fire or water the destroying element?   These and many other similar questions presented themselves for solution, but I had to confess my inability to arrive at a satisfactory answer to any of them, and thrust them aside for a more definite though not more interesting occupation, which, however, would be followed by certain tangible results.

On the twenty-third of January we started up the Rio Chico from our camp near Sierra Ventana.   We soon entered the great basalt-enclosed cañon through which the river flows for a distance of perhaps one hundred and fifty miles on its way from the Andes to the Atlantic.   Day after day we pursued the various meanderings of the stream, following the old Indian trail which for centuries had formed the chief highway of communication between the Indians of northern and southern Patagonia. Although now seldom used, the many paths worn deep into the surface of the valley, bore mute, but unimpeachable testimony as to the amount of travel that had frequented this highway in former times.   While the broken or discarded tolda poles and other accoutrements, scattered by the wayside, plainly bespoke the character of the travellers, by whom the trails had been worn throughout repeated journeyings across the unknown and uninhabited country, lying between the Rio Negro on the north and the Santa Cruz on the south.   At times our road would stretch for miles over the level, grass-covered valley.   Then for a short distance we would be compelled to resort to a dangerously narrow trail, skirting the side of some cliff as the river made a sharp bend and, encroaching upon the valley, rushed along over whirling rapids, or though deep quiet eddies at the foot of the escarpment beneath. At intervals, throughout each day during our journey up this cañon, we would halt for a few moments, or hours perhaps, to examine an exposure of sedimentary rocks projecting from the débris-covered slope beneath the basaltic platform, to collect specimens of each new plant as it appeared along our route, to take a new or rare species of bird or mammal, or to search the various springs and small brooks, that in certain localities were abundant throughout the valley, for fresh-water Mollusca, Crustacea, Planaria, and other forms of animal life, as well as such species of mosses, Hepaticæ, and other plants as are known to be partial to similiar habitats.

As night drew near, we would select a favorable locality by the bank of the river or some spring where there was an abundance of excellent grass for our horses and where dead calafate bushes were scattered about sufficient for fuel. Here we would encamp for the night and, after a splendid meal, prepared from our well selected stock of provisions and supplemented with the body of a duck, a brace of plover or other birds, or mammals, taken during the day, and carefully grilled over a bed of glowing embers, we would give such attention as was necessary to our collections and retire for the night.

For some time after entering this cañon we were thoroughly in harmony with and interested by our surroundings. The very novelty of the situation was pleasing. The valley of the river itself, by reason of its abundant food supply and the protection it affords from the frequent storms and almost constant winds of the high and level pampas, supports an animal life surpassing in diversity and numbers anything we had seen during our previous travels. Great herds of guanaco and flocks of ostriches appeared at frequent intervals. A variety of water-fowl frequented, not only the river, but the smaller springs and marshes along its borders. Among the reeds that grew in dense thickets in the shallow swamps, or fringed the borders of the smaller brooks, lived various wrens and other small birds including the delicate little flycatcher, *Cyanotis rubrigaster*, with orange-colored feet and a plumage presenting many shades of yellow, red and blue, blended in such perfect harmony as to rival in beauty and variety of colors those of the more tropical species of humming birds. From every bush and thicket the chestnut-crowned song sparrow, *Zonotrichia canicapilla*, could be heard from early morn until late at night, while, perched upon the highest branch of some particularly prominent bush, the Patagonian mocking bird, *Mimus patagonicus*, would at intervals take his stand and burst forth in joyous song with such sweet melody and in such volume as would quite fill the entire valley, and reverberate throughout the narrow defiles of the cañon walls. The repertoire of this little songster appeared unlimited. For hours at a time I have sat listening to one of these birds as, perched on a convenient bush, he poured forth at frequent intervals, but apparently quite at random, numerous strains from his seemingly inexhaustible supply, never repeating but always regaling his audience with something new. Flocks of the red-breasted meadow lark, *Trupialis militaris*, were common, while the black-throated, yellow-breasted sparrow, *Phrygilus melanoderus*, was everywhere. In the early morning from secluded places came the plaintive voice of the dove, *Zenaida auriculata*, so like the notes of that related species with which I had been familiar in my youth, that for an instant, forgetful of my surroundings, I fondly imagined myself transferred from our camp at the bottom of this deep cañon, more than a hundred miles from any habitation, to a comfortable bed under the paternal roof of my childhood and listening to the mournful sounds given forth by one of these birds perched safely on the limb of a tree above my window. The fond delusion, however, was but brief, for on thrusting my head from beneath the tarpaulin, I would discover not only that I was still in Patagonia, where in fact I was only too glad to be, but that daylight was at hand and with it had arrived the hour when we should be up and on our way. In the early evening a small burrowing owl, *Speotyto cunicularia*, might be seen flying about from one deserted burrow to another, while as darkness settled down, the yellow, or short-eared owl, *Asio accipitrinus*, and a gray species of similar size appeared in considerable numbers, as they glided in low, noiseless flight back and forth over the meadow lands of the valley in quest of such nocturnal rodents as might come within their grasp. And indeed there seemed no limit to this source of their food supply, for not only was the surface of the ground literally covered in many places with the well-used trails of these small mammals, which crossed each other in every conceivable direction, but in many places the earth beneath the surface was honeycombed to the depth of a foot or more with their subterranean burrows, in such manner that our horses sank at each step half way to their knees and the wheels of our cart plowed great ruts in the surface of the ground. Wherever there was a tract of warm sandy soil, numerous species of small lizards might be seen running swiftly about from one bush or stone to another.   The variety and beauty of color exhibited by these little animals was more striking than that of any of the other classes of animal life found in this region.   Of whatever hue and however brilliant the colors might be, they seemed to harmonize well with those of the surrounding objects so as to afford these comparatively helpless creatures a certain amount of protection from their natural enemies, by rendering them less conspicuous.   A specimen taken among the basaltic débris at the foot of the cliff would be nearly black with delicate spots of light gray arranged in transverse bands.   Another from a patch of mata verde, only a few rods distant, would be quite green, while a third captured on the almost barren surface of an alkali flat, or the bottom of a dried-up lake, would have the back and sides of body and head ornamented with light gray and delicate yellow scales arranged in broad vermicular bands on a dull brown, earth-colored groundwork.

After a few days spent in travel up the cañon we began to tire of our partial imprisonment and longed for a view out over the surrounding tablelands.   Owing to the tortuous course of the cañon we were seldom able to see for more than a few miles up or down the stream, while the high and precipitous nature of the cliffs shut off the view on either side.   The only direction in fact in which our vision could be said to be entirely unobstructed was straight up, and for our journey thither we were unfortunately as yet quite unprepared.   We pressed forward with eagerness to round each successive promontory, hoping to gain a more extended view beyond, only to be rewarded, however, with the sight of a very similar obstacle projecting into the valley in our front at a distance of only a few miles.   As we stopped for a couple of days to rest our tired horses and trap some of the rodents that were everywhere about us in great numbers, I decided to ascend to the summit of the basaltic platform on our left, in order to obtain a view of the surrounding country.   After a laborious climb among the sharp angular blocks I succeeded in reaching the top. Walking a few hundred feet across the fractured and cavernous surface of the lava to the top of a slight elevation, I stopped and carefully scanned the surrounding landscape.   Nothing could have been more desolate. Not even the great and silent wastes of polar ice so graphically described by Dr. Nansen could have presented a scene of more utter desolation than that which lay spread out before me. A black, barren waste of lava surrounded me on all sides. Its surface was distorted by huge overthrusts and indented by numerous deep, yawning chasms and caverns. Of life there was little or none. For indeed what animal or plant could find sustenance on such barren and inhospitable rocks? Notwithstanding the absolutely desolate nature of the scene about me, there was something both attractive and impressive in it, that led me to wander farther and farther over the surface, so that it was late at night when I retraced my steps and descended to our camp in the valley beneath.

About the middle of the forenoon of the second day after our halt we arrived at a point where the cañon makes a rather sharp turn and stretches away in an almost due westerly course, expanding into a broad, open valley or basin, which lies at the foot of the Andes. For the first time we now had a view of the forest-covered slopes and snow-clad peaks and summits of those magnificent and picturesque mountain ranges with which we were soon to become so familiar.

In the afternoon of this same day we came to the first of a series of great terminal moraines, which had been left at various stages in the upper course of the valley by a great glacier, as it receded from the plains. The position of this moraine is just above the forks of the river, where the Rio Belgrano enters the Rio Chico from the northwest. Our course lay up the Rio Chico, or south fork of the stream. On the north side, and immediately opposite this point, the high basaltic platform, which had followed the various meanderings of the valley, enclosing it like a mighty wall, with scarcely a break for a distance of a hundred miles, is interrupted by a broad, level plain, some fifteen or twenty miles in width. This stretches away to the northern horizon, between lofty basalt-covered tablelands, which enclose it on the west and east. At this point the old Indian trail, along which we had been travelling, crosses to the northern side of the stream, and, leaving the river valley, strikes due north over the surface of the plain just mentioned. With some difficulty we made our way through the rounded hillocks to the upper side of the moraine and camped near the river, just at the foot of a bluff of glacial silt. The side of the slope, as well as the little plain at its foot, was literally alive with rodents. Hopping about among the bushes and rocks, were to be seen in great numbers representatives of the little gray, tailless and hare-like *Cavia australis*. Most interesting and amusing little creatures they are, as, always alert and intent on detecting the first approach of danger, they hop about from one position to another, or sit erect on their haunches and nibble unceasingly at a fragment of plantain leaf, or other morsel of food held conveniently in the fore paws.   The favorite haunts of these little animals are shallow burrows about the bases of the larger bushes, or beneath certain herbaceous plants like *Bolax glebaria*, that grow in broad, dense, cæspitose masses upon the surface of the ground.   All about us, and indeed at times from immediately beneath our feet, could be heard the deep, subterranean drummings of the little tuco-tuco, *Ctenomys magellanica*, as engaged with commendable industry, he drove his little tunnel just beneath the surface, ever onward in search of those nutritious roots and succulent tubers upon which he feeds.   These little fossorial rodents seemed especially active in the early morning and late afternoon and evening.   During these hours, in localities especially favorable to them, they would be constantly heard, though a careful watch throughout our stay in Patagonia, kept at frequent intervals in order to observe their habits above ground, was only rewarded by a momentary glimpse, on one or two occasions, of a solitary individual, as he appeared for an instant at the mouth of a burrow.   On one occasion, however, while walking rapidly along, I came suddenly upon one of these little animals in the grass at a distance of several feet from the mouth of his burrow.   The manner in which he ran aimlessly about in search of his hole, with the nose close to the surface of the ground, seemed to indicate, not only that he had lost his way and become bewildered by the grass, which, to him, had all the appearance of a great forest, but that he depended quite as much, if not more, upon his sense of smell as that of sight, while endeavoring to regain the abandoned burrow.   Hardly had he entered the latter when the frightened condition under which he had been so evidently laboring while above ground, suddenly and completely disappeared, and he stopped long enough to send back a rapid volley of deep, guttural notes, uttered in defiance at the intruder, who, far from having cherished any sinister designs against the little creature, had only been delighted with this opportunity, brief though it was, of observing him above ground.   The entire attitude of the little animal was such as to convince me that his surroundings while above ground, aside from my presence, were distinctly uncongenial, and that he was in every respect especially modified and adapted for a subterranean life, a conclusion which I had previously reached upon observ-

ing the small eyes, powerful fore-limbs, and feet well adapted for burrowing, and other anatomical characters common to animals of more or less
subterranean habits.

In the small brush which grew at the base and over the slopes of the
bluff above our camp, there lived a variety of small rodents for the most
part characterized by large, thin ears, delicate soft fur of a bluish-brown
color above and lighter on the belly, with tails of various length, which in
some species might be described as short and in others much attenuated.

The tall grass which covered the river valley swarmed with myriads of
small rodents of other species somewhat larger than those just mentioned,
with usually smaller ears, shorter tails, and a coarser pelage of an almost
uniformly dull brown color. While these little animals were present in
the greatest abundance, they seemed all to pertain to one of two or three
different species and exhibited very little variety of either form, size or
color.

During the first night passed at this camp by the river above the first
terminal moraine, we experienced the most severe rain storm witnessed
during our travels in Patagonia. It commenced about ten o'clock and
continued throughout the greater part of the night. The rain fell in perfect torrents, while there was an almost continuous roar of thunder accompanied by most vivid flashes of lightning, which followed one another in
rapid succession. We had received ample warning of the approaching
storm the evening before and had so prepared for it that little or no
injury resulted to either ourselves or our belongings. On walking about
the following morning I was much impressed, not only with the amount
of erosion which had been effected by the storm, but with the great destruction to animal life of which it had been the cause. Proceeding along the
foot of the bluff, I observed a number of dead bodies of the little *Ctenomys
magellanica* lying upon the surface, and could only guess at the number
of carcasses of these and other rodents that were buried beneath the débris,
that during the night had been washed down from the bluff above and
now lay at my feet, covering no inconsiderable portion of the surface of
the valley to a depth varying from an inch to one or two feet. While the
dead bodies of rodents belonging to other species were not wanting, the
storm appeared to have been especially destructive to the little tuco-tuco,
owing no doubt to the peculiar habit of that animal in burrowing so near
the surface of the ground in search of food. When erosion on the surface

had made an opening in a burrow at any point, a torrent of water would rush into the subterranean channel, either instantly drowning such of its inhabitants as were caught below, or driving them to seek refuge by escaping from the burrow where they were certain to meet with a similar death from the downpour of rain on the outside. As I walked about this morning, considering the destruction wrought by the storm of the previous night, I was struck with the great importance of the work accomplished by rodents and other burrowing animals, when considered as agents of erosion, and it appeared to me that this source of erosion had not been given sufficient attention in our text-books of geology, when treating of the various erosive agents.

The storm had rendered the surface of the surrounding country quite unfit for travel, so that we were compelled, not unwillingly, to remain where we were for the day, and indeed there was much of profit to be gained from our sojourn. Mr. Peterson found ample employment throughout the day in caring for the supply of rodents taken in our traps during the night, which had been supplemented by others picked up on my walk in the early morning. Personally I spent the forenoon in examining the bluff a short distance below camp, where the river had cut a narrow channel through the accumulation of silt and other materials, which had been brought down and deposited by the ancient glacier in such manner as to obstruct the natural course of the river. The materials of the bluff were found to consist for the most part of fine silt. This exhibited splendid examples of a peculiar, complicated, irregular, concentric structure which had been wrought out in high relief by wind erosion, from which I secured the photograph reproduced in Fig. 13. In the afternoon I saddled a horse, in order to find a safe place at which to ford the river with our cart, since the travelling for a considerable distance seemed much better on the opposite side. Having successfully accomplished this purpose, I proceeded to the summit of some high hills that lay at a distance of several miles to the northwest and formed the divide between the Chico and the Belgrano. From the top of one of the higher of these I secured an excellent view of the surrounding country. Between the western limits of the high basaltic table-land of the interior and the eastern foot-hills of the Andes, there lay a wide tract of open country, well grassed and for the most part free from lava. A few miles farther up I could see that the river forked again. The southernmost of these two branches came from

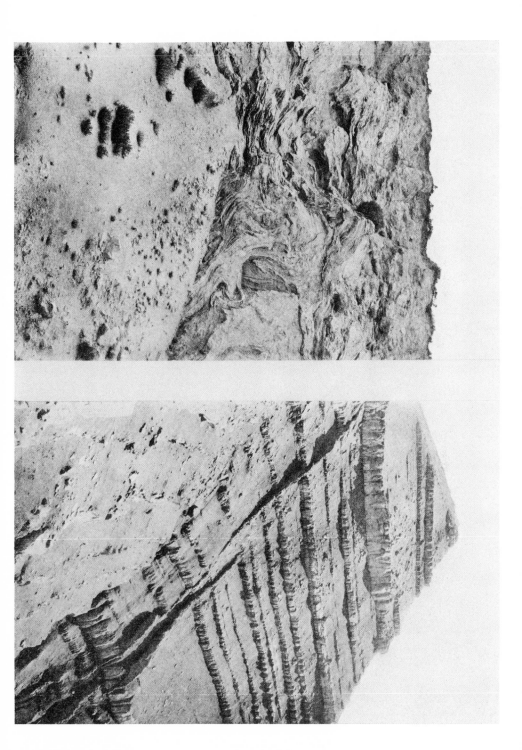

Figures 13 and 14—See other side

a large circular basin, which lay just at the foot of the Andes and was partially surrounded, especially on the south, by bare, white, bad-land hills. The extensive and promising nature of these exposures led me to decide upon entering the Cordilleras by way of the southernmost of these two branches of the river. After examining some small exposures of sandstone on the side of the hill at the summit of which I had been standing, without arriving at any very satisfactory conclusion as to their exact nature, I returned to camp and the following morning we resumed our journey up the river toward the foot of the mountains.

We crossed the river almost immediately and continued to travel throughout the day on the north side, camping at night alongside a beautiful spring of clear water which came boiling up from a bed of almost pure white sand at the bottom of a deep hole, some ten or twelve feet in diameter, and flowed away across the meadow land which lay between us and the river, in a brook of no inconsiderable dimensions. Hardly had we halted and began unhitching our horses from the cart than a carrancha came and perched himself at the top of a calafate bush so aggressively near that, out of sheer wantonness, I am forced to admit, I drew my six-shooter from its scabbard and, just to keep my hand in, dropped him lifeless to the ground beneath.

The following morning we resumed our journey at an early hour and about the middle of the forenoon recrossed the stream some distance above the forks last mentioned and continued our westerly course along the north bank of the more southern of these two forks. Since leaving the old Indian trail the nature, both of the valley and its surface, had completely changed. Instead of the deep and narrow basaltic cañon with a perfectly level bottom covered with loose sand and shingle, there was a broad, open valley with low, rounded glacial hillocks interspersed between extensive tracts of bright green meadow lands, dotted over with small glacial lakes and surrounded by high rolling hills.

Long after noon we arrived at a narrow gap cut by the stream in a low ridge of sandstone belonging to the Patagonian formation. So narrow was this gap that we had to take to the bed of the stream, in order to pass through it with our cart to the broad open basin that lay beyond. Since the surface of the valley over which we had been travelling during the early morning had been very rough, we decided to give our horses a short rest and spend the time ourselves examining the rocks constituting the

valley of the gap through which we had passed.   As just stated, we found
these to belong to the Patagonian formation.   They were chiefly sandstones and were inclined at a high angle to the eastward.   They were rich
in the remains of marine invertebrates.   A species of oyster, *Ostrea ingens*,
common to that formation, was especially abundant and of exceptional size
even for that unusually large form.   Frequent examples were observed in
which a single valve would measure a foot in length and the combined
weight of a single specimen, including both upper and lower valves, could
scarcely have fallen much, if any, short of forty pounds.

After resting and grazing our horses for an hour or two at this place,
which we christened Shell Gap, we proceeded a few miles farther to a
thicket of small beech trees or bushes growing at the foot of a low hill,
some three or four hundred yards beyond where the stream flows through
a rugged cañon, cut in the southern walls of the great valley which we
named Mayer Basin.   Here, convenient to a spring of splendid water
with an abundance of grass for our horses and firewood sufficient for our
purpose, we proceeded to establish a permanent camp in a well sheltered
place, where we might spend a number of days in making collections representative of the surrounding flora and fauna and in studying the geology
and geography of the vicinity.   Hardly had we pitched our tent, put things
to rights and set a few dozen traps in apparently advantageous places,
when a cold drizzling rain set in and continued throughout the night.

CHAPTER VIII.

An outlier of the forests; Cretaceous and Tertiary rocks at the foot of the Andes; Abundance of petrified wood; Cariacus chilensis; Cross the continental divide; Enter the Andean forests; Western portion of Mayer Basin; Discover Mayer River; Mud streams, Dinosaur bones and Alpine plants on high bluffs south of Mayer Basin; Numerous glaciers; Searching the Dinosaur beds for mammalian remains; Life about and in the forests; The gray-banded woodpecker; Habits of the little green parrakeet; Avian life in the depths of the forest; Grebe in the small mountain lakes; Fearlessness exhibited by the deer of this region; The barking bird; Bald Mountain; Glaciers in Mayer Basin; Outskirts of the Andean forests; A trip to the Glaciers; Characters of the forests.

ON the following morning I set out at an early hour to examine a bad-land exposure, which lay at a distance of some two miles in a southwesterly direction from camp, Mr. Peterson remaining in camp to attend to the rodents that had been taken in our traps during the previous night. Between our tent and the exposure mentioned above, a level valley extended to the edge of a small, wooded tract, which lay in front and at the foot of the bad-lands. With much impatience I trudged along through the grass, which was thoroughly soaked with moisture and reached half way to my hips. At every step I sank nearly to my knees in the half decayed vegetation, which had been accumulating for years on the surface. The latter was extremely uneven, owing to the numerous burrows and paths that had been made by the thousands of small rodents inhabiting the region. I have seldom experienced more difficult walking and was quite disgusted, when, thoroughly saturated quite to my waist, I arrived at the edge of the little forest. However, I soon forgot my discomforts and became lost in the interest and novelty attending this my first experience in a primeval Patagonian forest, for, during my enforced stop at Sandy Point, the short excursions made into the woods back of the town hardly brought me without the influence of human habitations, and could in no case be considered as affording any adequate idea of the bird and mammal life indigenous to the forests of the southern Andes. Although the dark and gloomy nature of the day was such as would exert a particularly depressing influence among the birds, hardly had I entered the forest when the incessant twittering of the beautiful little Chilian wren, *Troglodytes hornensis*, could be heard all about me. As I stopped for a moment these little birds, with habits more closely resembling those of the creepers than of the wrens, appeared in great numbers, hopping about on the branches and trunks of the trees, perfectly fearless, approaching to within a few inches of my outstretched hand and exhibiting a degree of confidence in human nature previously altogether unobserved by me in any species of wild bird. In the upper branches of the trees might be heard almost constantly, and occasionally seen, a small white-crested fly-catcher, *Elainea albiceps*, which was as consistently shy and unapproachable as the little wren just mentioned was tame. Two or three species of flickers were also not uncommon.

It took but a few moments to cross through the narrow wooded strip to the bad-land hills beyond. These I soon found to consist, for the most part, at this point at least, of light ash-colored materials, with an abundance of lignitized and petrified trunks of trees, but singularly destitute of animal remains. It is difficult for one who has not seen it to imagine the abundance of fossil wood in these beds. In one locality I came upon a place in the side of a cañon, where, for a depth of perhaps forty feet, the walls on either side were entirely made up of the petrified trunks and branches of trees. Some of the former were several feet in diameter and many feet in length, and they were interlocked one with another in such manner as to suggest that originally they had formed a natural dam in the current of some prehistoric stream.

After a few hours spent in a vain search for mammalian remains in these beds I started for camp, returning by way of the forest through which I had passed on my way thither. Just as I was emerging from the wooded tract into the meadow land in front, I came suddenly and unexpectedly upon three deer browsing quietly in the grass along the margin of the wood. They were the first I had seen in Patagonia, and for a moment it was evident that I was the most startled individual of the four. They made no effort to escape, as they might easily have done by taking to the wood, but stood at a distance of not more than twenty feet, returning my expression of surprise with one of interested curiosity. For an instant I stood admiring the rich golden brown of their sleek, glossy coats, as they alternately cropped the rosebuds and other choice morsels from the foliage about them, or cast inquiring glances toward me. Suddenly, remembering that we had been without fresh meat for breakfast, I deliberately, though reluctantly, drew my revolver from its scabbard, and having for a moment subdued the compassionate feeling with which I had been seized, it required little skill to despatch one of the trio and demonstrate that man is not less brutal than other animals. Indeed, from a certain, and to my mind questionable standpoint, it was about as unsportsmanlike an act as could have been committed. But, like others even more unsportsmanlike which I shall later have occasion to relate, it served the double purpose of providing us with a supply of meat and an addition to our collection of the skins of recent Mammalia. The two survivors remained, unalarmed either by the report of the fatal shot, or the death struggles of their companion. While I was engaged in skinning and dressing the carcass of the dead animal, the live ones stood at a distance of only a few yards, either indifferent, or at most only curious as to the nature of the operation, and I could at any moment have easily despatched them, had I been so inclined. Covering the carcass and skin with brush so as to protect them from the carranchas, I returned to camp and, saddling a horse, conveyed them to our tent, where they were properly cared for.

We remained at this camp for a few days, examining the adjoining badlands for fossils and making no inconsiderable additions to our collections of recent birds, mammals and plants. After learning what we could of our immediate surroundings, we crossed over the low continental divide and camped on a small stream, which flows toward the Pacific along the western border of the valley which we had already designated as Mayer Basin, in honor of our friend General Edelmiro Mayer, Governor of the Territory of Santa Cruz. It is a singular fact that, except in the region of Lakes Viedma and Argentino, the main ranges of the southern Andes nowhere form the watershed between the Atlantic and Pacific Oceans. Throughout all the remainder of that vast extent of country the continental divide or watershed lies far to the east of the main ranges of the Andes. This fact, which remained for a long time unknown, has been the source of considerable friction between the Argentine and Chilian governments, when attempting to fix the boundary limits between the two countries.

Our camp on this small stream lay just in front of a lofty and precipitous bluff of hard, coarse standstones, which formed the southern border of the western half of Mayer Basin. We were now on the very edge of the dense forests that everywhere clothe the lower slopes of the southern Andes. To the south and southeast were considerable exposures of sedimentary rocks, that as yet remained to be examined. The bluff which formed the southern border of Mayer Basin, directly south of camp, had an elevation of perhaps three thousand feet above the stream and its summit afforded an exceptionally fine view of the surrounding country. On the morning following our arrival at this camp, I climbed to the top, examining the different strata as I passed over them. The coarser sandstones at the bottom were quite barren of fossils and seemed very similar to those we had previously seen near the source of the River Chalia, or Sheuen. Here, however, they were overlaid by several hundred feet of variegated clays, alternating with an occasional thin stratum of sandstone. At several localities in the clays I found fragments of the bones of dinosaurs, and occasionally a nearly complete vertebra, or footbone. In a pinkish colored stratum near the summit I came upon a nearly complete fore limb of a large dinosaur. The weight of the humerus alone could hardly have been less than two hundred pounds. I wished very much to take this limb, but its great size and weight would have precluded our taking it with us to the coast, hence I left it where it was, hoping that I might again return to the same locality on some later expedition, better equipped with means of transportation.

The face of the cliff was perfectly bare of vegetation, and where the clays predominated, the slopes were exceedingly slippery from being thoroughly saturated with moisture by the recent rains. This rendered their ascent exceedingly difficult and not wholly without danger. I fear my poor horse had arrived at the conclusion, long before we reached the summit, that he had fallen into the hands of a hard master. What rendered the travelling even more difficult was the presence of great numbers of mud-streams encountered at the bottoms of the smaller gulches, with which the face of the bluff was furrowed. Usually these were only small affairs and could be easily crossed, but in not a few cases they were of no inconsiderable dimensions, being several rods in breadth and of unknown depth. Moreover, the surfaces of the larger of these were not always of a character to reveal their true nature. On the top there would be formed a half-hardened crust, rigid enough to support bowlders of considerable dimensions. The presence of these, together with the generally level appearance of the surface, would make it seem to offer a safe highway much to be preferred to the steep and slippery slopes on either side. Hardly would the poor horse leave the difficult incline and take to this comparatively level highway, when he would break through the treacherous crust and find himself floundering in several feet of thick, sticky mud, from which he was frequently quite unable to extricate himself, until after being unsaddled, and then only with the greatest difficulty. These streams of mud were phenomena previously quite unknown to me, but it needed only a few experiences with them to teach me to recognize and avoid them. I afterwards observed that they were actual mud-streams, commencing in many instances near or at the summit and gradually augmenting in volume, by the accession of occasional tributaries, until, at the mouth of the main gulch, they spread out in a broad, fan-shaped mass upon the surface of the valley like the termination of a miniature glacier.

After a long climb, attended by many interruptions, I reached the summit, on the surface of which I found growing a few scattered Alpine plants· For the most part, however, it was quite bare. From my elevated position I could look off over the surface of the great, black, basalt-covered plains, which lay to the east. Looking to the northward, down the valley of the little stream on which we were camped, I could see between the mountains in the distance, the broad, deep valley of a great river which I knew had not been previously discovered. Beyond this lay a broad, circular basin, enclosed by the most rugged and picturesque mountains I had ever seen. Through the deep cañons, eroded in their precipitous slopes, numerous glaciers descended from their snow-clad summits to the basin below, while a considerable river flowed from the front of each of these glaciers out across the valley to unite its waters with those of the larger stream, which I had already decided to call Mayer River in honor of General Mayer.

To the south the ridge expanded into a series of bare, rounded hills or hogbacks. These had a general easterly slope and gave rise to a number of small brooks, which united below to form that fork of the Rio Chico by which we had entered Mayer Basin through Shell Gap. The surface was, for the most part, swampy and on the lower slopes a low trailing shrub,

Pernettya empetrifolia, completely covered the surface of the ground. This in turn was literally buried beneath a mass of red berries, about the size of the small red currant, and scarcely inferior to the latter in flavor. These berry patches, which in places actually covered the entire surface, were at this season of the year favorite feeding grounds for certain species of large grouse-like plover, while ducks, geese, and other waterfowl were not infrequently seen partaking of the rich, juicy fruit. Under cultivation, I see no reason why so hardy a plant, so prolific in the production of fruit of such excellent flavor, might not become of considerable economic value. I dried and brought with me to the States a quantity of the berries and seeds, which were distributed among certain cranberry growers in New Jersey, beside sending a considerable quantity to the Agricultural Department at Washington, in hopes that they might be successfully grown and prove of some economic value. Having heard nothing from them, I infer that in either instance they were so carefully or carelessly put away for safe keeping, that they have not been since resurrected and transferred to the soil.

I passed the day on the summit of this ridge, which was somewhat above timber line, and, although the month was February, there still remained occasional snow drifts. Of plant life there was little, while of animals there were almost none, save the plover and waterfowl already mentioned. Such plants as there were, were to a great extent limited to a number of hardy lichens clinging tenaciously to the rocks and a small variety of subalpine flowering plants, which, for the most part, I had not seen elsewhere. Among others there was a species of *Oxalis* with such delicately dissected leaves and exquisitely beautiful pink blossoms that it seemed strangely misplaced and out of harmony with so apparently uncongenial surroundings. Descending to the lower slopes of the head of the Rio Chico, I met with a variety of Compositæ belonging, principally, to the genus of asters, which for abundance, beauty and variety of colors, I do not believe can be excelled elsewhere by representatives of this family in a state of nature and entirely unaided by artificial cultivation and cross-fertilization. Despite the terrific wind which prevailed and was accompanied at frequent intervals by snow, rain, hail and sleet, I passed a most interesting and profitable day on the summit, and returned in the evening to our camp in the valley with a considerable number of most welcome additions to our botanical collections. Descending from the crest of the bluff to the river was like passing from early spring into mid-summer. The sun was still shining and a glow of warmth pervaded every-thing. Mr. Peterson was much surprised when I told him of my experi-ence with the weather. He assured me that in the valley the day had been one of the most pleasant he had yet experienced in Patagonia. For the first week or ten days of our stay at this camp I visited daily the sum-mit of the bluff just described, in a vain search for remains of that won-derful Pyrotherium mammalian fauna which, according to Ameghino, should be found here in association with remains of dinosaurs. Not a day passed that I did not find remains of dinosaurs, but never the smallest fragment of a mammal. The weather conditions were always the same, alternate sunshine, snow, rain, hail and sleet, swept by a cold, piercing blast that never ceased to come from out the snow-clad peaks and ranges of the Andes. Finally, discouraged with my search after fossils, I aban-doned it and took to the more pleasant occupation of assisting Mr. Peter-son with his work in connection with the recent birds and mammals.

Each morning, after attending the traps set the previous night for mam-mals, I would take the fowling piece and, with auxiliary barrel and a num-ber of assorted cartridges, set out along the borders of the great forest near which we were encamped. In the early morning the edge of the wood is always much better collecting ground than are the dense forests within. Here, feeding upon the abundant insect life that always frequents such places, were to be found a number of small wrens and sparrows, the white-crested bunting, flocks of the red-breasted meadow lark, the Pata-gonian mocking bird and a number of thrushes and flickers, including the gray-banded woodpecker, *Colaptes agricola*. These birds were usually observed in small flocks of four or five, either clinging to the trunk, or sitting in the branches, or on the ground at the base of the half decayed stump of a tree still standing in the grass just without the edge of the forest.

After a couple of hours spent in the vicinity of camp I would return with such game as had been taken and, replenishing my supply of ammunition, strike out to spend the day in the midst of the great forest in search of such birds as were only to be found within its depths. The little green parrakeet, *Cyanolyseus patagonicus*, was much the most abundant of the birds frequent-ing the interior of the forests. In places they occurred absolutely by hun-dreds, if not thousands, and were quite distracting by reason of the harsh chattering notes which they emitted continually from the tops of the trees, as one walked along underneath.   If, for any reason, one discharged a gun while in the near vicinity of a flock of these birds, they would instantly break forth into such a pandemonium of noises, as might easily lead one to think that all the furies of the nether region had been turned loose.   At certain hours of the day, however, they were exceptionally quiet and one would be cognizant of their presence as he walked about over the moss-covered surface beneath the trees, only from a faint chattering emitted at intervals by a few scattered individuals.   On pausing for a moment to survey the tops of the surrounding trees, a few of these birds might perhaps be seen clinging to the under surfaces of some of the lower of the wide-spreading branches overhead.   But the most minute and careful inspection would by no means reveal their true numbers, so perfectly did the green color of their feathers harmonize with that of the leaves of the branches about them.   If, however, one was selected as a target and a volley of number eight shot discharged, it would frequently result in the destruction of the lives of several other birds beside the one fired at and create a scene of the greatest activity in the surrounding forest, where but a moment before there had appeared scarcely any evidence of animal life. In a few moments, however, they would gradually settle down into their former quiet state.

Climbing about over the trunks of the trees were numerous examples of the large brown creeper, *Pygarrhicus albigularis*, assisted by its strong bill and forked tail terminating on either side in a few, stiff, pointed feathers, while the little black *Scytalopus magellanicus*, with habits very similar to those of the wrens, could be seen hopping about alternately on the ground and among the branches of the underbrush, all the while in such constant motion that it was exceedingly difficult to obtain more than a momentary glimpse of the tiny creature.

Occasionally the loud, clear, resonant notes of the magnificent red-headed woodpecker, *Ipocrantor magellanicus*, would be heard in the distance ringing through the silent depths of the forest.   This beautiful bird, its head ornamented by a crest of brilliant red and body clothed with a covering of glossy, coal black feathers, is one of the most striking and at the same time least common of the birds of this region.   Moreover, as it was exceedingly difficult of approach, we succeeded in securing only a single specimen.

Of owls also there was a considerable variety, we having taken no less than five species at this one camp.

One particularly bright and cheerful day late in February, as I rode through the woods at a distance of some five miles west of our camp, I came upon a small, nearly circular lake of about one mile and a half in diameter. As I emerged from the forest and sat on my horse by the rocky shore, where I thought to stop for a moment and admire the beautiful scene before me, there came floating across the water from the far side of the lake a low plaintive sound, which I instantly recognized as that of the grebe, *Æchmophorus major*. In this sheltered place there was not a sufficient breeze to cause the slightest ripple on the surface of the lake, which for an instant I carefully scanned, hoping to get sight of the flightless bird which I knew must be present, though the locality was remote from its normal habitat. For a few moments, save the low, plaintive cry which was wafted at intervals from the opposite side, I could see nowhere on the surface of the lake the slightest evidence of life. A little later, however, I detected a wide V-shaped ripple on the water, with a small black object at the apex which was directed straight toward me from the opposite shore. For a time I remained motionless and watched the solitary bird as he sat gracefully on the surface of the water, with his long neck erect, and held a perfectly straight course for the beach at my feet, continuing to utter at regular intervals those singularly plaintive notes which seemed almost as though intended to bespeak from me commiseration for him in the lonely solitude of his surroundings. Still onward he came with supreme confidence, until he had arrived at a reasonable distance from the shore, when I levelled my fowling piece and dropped him lifeless on the surface of the water, just as his last uttered note was losing itself in the depths of the forest at my back.

In the early morning and late afternoon deer were common about the edges of the wood and in the small open parks within, while in the middle of the day they were frequently met with in the depths of the forests. The degree of confidence and fearlessness displayed by these traditionally timid animals was indeed most remarkable. It was plainly evident that they were entirely unacquainted with man. On one occasion, while tramping through the woods with my shotgun in quest of smaller game, I came upon a full grown male lying quietly at the base of a large tree. As I stopped to observe him, he remained quite still for a moment and looked at me, with nothing of fright in his countenance. Then slowly getting upon his feet he came walking directly toward me with that measured and firm tread characteristic of the family. The entire attitude and bearing of the animal resembled those of a favorite cow or horse, as, lazily basking in the barn-yard, it rises and advances slowly to lick the proffered hand of its master. I permitted this exhibition of confidence to continue until he had approached to within some ten or twelve feet of me, when I showed my unworthiness by exchanging a charge of small for one of solid shot, which, after retreating for a few paces, I discharged with such effect that the beautiful animal fell lifeless almost at my feet, a victim of misplaced confidence.

After several days in this camp, passed in much the same manner as that just described, sometimes varying my hunting expeditions with excursions after mosses, Hepaticæ, flowering plants, or other botanical materials, while Mr. Peterson was busily engaged with the rodents, of which he secured a splendid collection, including a fine series of a much larger species of *Ctenomys* than any we had seen in the plains regions, we moved some ten miles farther down the stream and camped by a little brook at the edge of a forest, which covered the lower slopes of a considerable mountain, lying to the north of the valley of the stream. Between our camp and the stream lay a broad stretch of meadow land covered with an abundant growth of grass, which was the home of innumerable rodents. In the evening this meadow became the feeding ground for great numbers of owls. By hanging a lantern on the limb of a tree which stood in front of our tent, or building a small camp fire, these birds were attracted by the light and frequently brought within range of our fowling piece, so that we secured at this camp alone a very fine series of no less than five species. Near the brook within the forest by which we were camped, there lived a considerable variety of small rodents. It was a part of my occupation each day to attend a number of traps placed at frequent intervals and in what appeared to be particularly promising localities along this little water course, in hopes of procuring in the interior of the forest a number of species of rodents different from those which frequented the meadow lands and edges of the woods. In this I was only partially successful. While attending these traps one early morning, I was suddenly attracted by a loud barking noise, emitted at regular intervals, which came from the deep forest directly in front of me. The notes were in every particular very similar to those of some of the smaller varieties of the domestic dog, and for an instant I thought I was in a fair way of securing a specimen of *Canis magellanicus*, the large reddish timber dog of this region, examples of which we had not as yet seen. As I stood for a moment carefully scanning the openings between the trees, hoping to catch a glimpse of the object from which the sounds came, I noticed that with each successive note it drew nearer, and presently I saw emerging from a dense growth of underbrush at a little distance in front of me a beautiful example of *Hylactes tarnii*, the barking bird of these Andean forests. The little creature came hopping along the ground directly toward me. The beautiful, plump breast, covered with delicately mottled chestnut brown feathers, rose and fell successively, and at each stage of his advance he gave forth a note singularly like that of a small dog. I retreated before the advancing bird to a distance sufficiently remote to insure it from being blown to atoms and reluctantly despatched it with a charge of the finest shot with which at that moment I happened to be provided. Although this bird is said to be fairly common in the forests to the north, it proved to be the only example we met with throughout our travels in Patagonia.

After a few days passed in this camp, we moved on down the stream for a distance of ten or fifteen miles, encamping a short distance above a considerable cataract, where the nature of the country was such as would have rendered our further descent with the cart quite difficult, if not impossible. A few miles below our last camp the stream received from the south a tributary about equalling it in size. Immediately below this the valley became narrow and the water flowed along over a bed of shingle, or between huge bowlders in an extremely tortuous course. This compelled us to cross and recross the channel at various intervals during our descent, as it swung back and forth from one side to another of the enclosing bluffs. Travel, therefore, became somewhat difficult for several miles before we decided on camping above the waterfall just mentioned.

In this camp we remained until the first of March, giving our time almost exclusively to the collecting of the skins and skeletons of recent birds and mammals and in making additions to our botanical collections. We were now in about the center of Mayer Basin. On our north there rose a considerable elevation which we had christened Bald Mountain from the nature of its summit, which though above timber line, was not sufficiently high to be covered with perpetual snow.   Bald Mountain is an isolated dome standing in about the middle of Mayer Basin.

The western face of this mountain afforded an excellent view of the snow-clad peaks and ranges of the higher Andes, and of the numerous glaciers that descended the slopes of the latter on the western border of Mayer Basin.   At the time of our visit each of these glaciers descended far down among the forest-covered slopes, and from the front of each there poured forth a considerable river.   These flowed out at first in an easterly direction, emptying their waters into Mayer River, which swept away to the westward through a narrow, impenetrable mountain defile, by which it enters the Pacific.   In one of these glaciers on the extreme left of the basin there is a grand cataract, where the ice plunges over a mighty precipice several hundred feet in height.   The different shades of blue, green and white displayed by the ice in the front of this were exceptionally beautiful.

On a particularly fine day late in February I laboriously climbed to a position above timber line on the western end of Bald Mountain.   Here, with an entirely unobstructed view, I sat for some time enjoying the magnificent panorama which lay before me.   The great river rolled swiftly on through the valley below.   Beyond this lay the dark green forests of beech which covered the basin and lower slopes of the mountains.   In places the foliage of the forests was already tinged with yellow, purple, red and other autumnal colors, while beyond and above the whole towered the magnificently rugged central range of the Andes, buried beneath enormous fields of snow and ice, which covered all as with a brilliantly white mantle, save at intervals, where some particularly bold promontory or sharp and jagged peak raised its giant form like a black sentinel high above the surrounding fields of white.

After some time passed in enjoying the splendid view which my position afforded, I resumed my journey and continued the climb to the summit of the mountain.   I then walked along the crest until I came to a spot directly opposite our camp, when I decided to descend the mountain slope and thus make my way to camp.   So long as I was above timber line I succeeded very well, but when I approached the outskirts of the forest, my progress was seriously interfered with by such a tangled growth of low spreading beech bushes as made travelling well-nigh impossible.   The trunks of these bushes never grew erect, but wound and twisted about at

FIGURES 15 AND 16—SEE OTHER SIDE

15—In the forests near Sandy Point.

16—Mayer Glacier.

a distance of a few feet or inches above the surface like huge vines, sending upward such a profusion of little short, leafy branches as rendered it impossible to see for more than a few inches beneath. It was absolutely impossible to proceed by walking on the ground, so that for a distance of a mile or more I literally walked along over the tops of the woods, until I reached the interior of the main forest, where the trees were erect and unobstructed, I then made rapid progress down the heavily wooded slope to camp. I found on the upper and lower outskirts of all these beech forests very similar conditions to those just described, and I believe it due, to some extent at least, to the deep snows that must accumulate here in winter, by which the young trees are broken and bent down with the weight of snow from year to year and thus prevented from accomplishing a natural growth.

The little stream on which we were camped flowed for several miles below us through a rather deep and picturesque mountain cañon, before entering the main valley of Mayer River. This was a favorite resort for several birds not common elsewhere. In one particularly rocky defile, where the stream went meandering down among great numbers of huge bowlders, I secured a specimen of *Nycticorax cyanocephalus*, a night heron with two delicate, pure white, thread-like plumes on the occiput, some six inches in length. On another occasion I came upon a small flock of chestnut colored geese, *Chloëphaga poliocephala*, feeding in an old deserted channel of the stream. From among these I secured our only specimen of this goose.

Having decided not to proceed farther into the Andes with our cart, on March 1, we made all secure about camp and, taking sufficient bedding and provisions, set out with improvised pack saddles on an expedition to the glaciers beyond Mayer River. We were not long in selecting and packing the few necessary articles with which we could not dispense and were soon on our way. Having converted our cart-horses into pack animals, and mounted on our saddle-horses, we had a small but exceedingly mobile and efficient pack train quite sufficient for our needs. We made our way through the forests along the stream on which we were camped, until we came to the main branch of Mayer River, at a point a little distance above where the first of the rivers flowing from the glaciers in Mayer Basin emptied into that stream. At this place the bed of the river is perhaps five hundred yards in width, but with the stage of water then in the stream the current was broken up into a great number of channels separated by low shingle beds, so that we had no difficulty in effecting a crossing.   After crossing the river we emerged upon a low, broad and level valley lying between the river and the forest.   As we crossed the grass-covered valley and entered the edge of the forest, we came upon a deer, and since we had not provided ourselves with meat, having depended, as was always our custom when travelling in Patagonia, upon such game as should come to hand, we stopped long enough to slaughter it and dress the carcass.   After caring for the skin, skull and such limb bones as would be necessary for properly mounting the specimen, and placing them where they would be secure until our return, we kept on our way through the forest, until we arrived, late in the afternoon, on the banks of a river, the southernmost of those that flow from the glaciers in Mayer Basin, at a point just below where it emerges from a rather deep gorge, and as we subsequently ascertained, some three or four miles below the glacier in which it has its source.   The stream was not fordable, so we camped for the night, and the following morning, leaving our pack horses picketed on both water and grass, so that they would not suffer in case we were not able to return, or chose to remain away for two or three days, we took our saddle-horses, camera, one rifle and such bedding and provisions as were absolutely necessary, and proceeded up the river by the south bank to the glacier.   It proved to be literally alpine climbing on horseback and was incomparably the worst country over which we had ever taken horses. It is a thousand wonders that our horses ever succeeded in getting through alive.   However, we accomplished the journey without any serious injury to either ourselves, our horses, or outfit, and arrived at the front of the glacier at about ten o'clock.   The day was everything that could be desired.   Not a cloud could be seen and, as we rode upon the top of the terminal moraine and looked at the great, white river of ice from three to five miles in width and perhaps forty in length, I thought it one of the most beautiful and impressive sights I had ever seen.   I had never before seen a glacier, except at a distance, so that the entire situation was both novel and interesting.   The peculiar rounded hillocks of the terminal moraine were most instructive and explained many phenomena which I had noticed on the plains.   The terminal moraine at the time of our visit was several feet higher than the front of the glacier, and this in turn was somewhat higher than was the surface of the ice at a distance of a quarter

FIGURES 17 AND 18—SEE OTHER SIDE

17 – MAYER GLACIER.

18 – SHEEP FARM, NORTH GUER AIKE.

or half mile up the glacier. We should have liked to explore the surface of the ice, but unfortunately a river from another glacier and lake, which lay above, entered this river just between the moraine and the end of the glacier, which prevented us from reaching the surface of the latter. We passed the remainder of the day in photographing the various aspects of the glacier and moraine, and in making such observations as were possible. Two of these photographs are reproduced in Figs. 16 and 17. A small bird which we had not noticed elsewhere was abundant, and we regretted that we had not brought a shotgun rather than a rifle. We had selected the latter weapon, hoping that we might happen upon one of the larger of the two species of deer which are said by some to inhabit this part of the Andean forests. Of this animal, however, we never saw any sign, and I seriously question whether it ranges so far south. Except for the bird just mentioned and a few insects, there was little of animal life to be seen. The glacier terminated in the midst of the upper limits of the forest and immediately below the moraine there was a splendid clump of trees. In the midst of these we built a rousing camp fire, around which we sat, until far in the night, talking alternately of our present surroundings and our experiences of the past few months, with occasional conjectures as to the health and doings of our relatives and friends at home. At last, wearied with the exertions of the day, we spread our blankets on the ground alongside the still glowing embers of our nearly burned-out fire, and, with our saddles for pillows, retired for the night and were soon lost to our surroundings in a deep, refreshing sleep.

On the following day we returned to our camp down the river, which we gained with scarcely less difficulty than we had experienced on the previous day in ascending to the glacier. After our return the remainder of the day was passed in collecting a few imperfectly preserved Ammonites from a bluff of slate on the river bank, at the mouth of the gorge a short distance above camp, and in making some additions to our botanical collections. In the evening Mr. Peterson killed a deer which entered our camp in such an impudent manner, as might lead one to believe that he intended taking full possession.

On the following morning we started on the return to our main camp. Since we had experienced some difficulty in handling three inexperienced pack horses in the thick woods that lay between us and Mayer River, we decided to pack all our belongings on two horses and let the third horse follow, while we would each lead one of the packs.   This plan worked for a short distance only, for hardly had we got well started on our way, when the loose horse, falling a little in the rear, became bewildered and ran squealing about through the dense forest in search of his companions, making all the while so much noise himself that he could not hear the answering calls of ourselves or our horses, and finally became entirely lost to either sight or hearing.   Giving his pack-horse to me, Mr. Peterson set out in search of the lost animal, while I resumed my journey.   Before reaching the place where we had previously killed the deer, on entering the woods after crossing Mayer River, I came upon another which I also shot and skinned.   When I went to get the skin of the deer first killed, I found that Mr. Peterson had been there before me and had left a note to inform me that he had not found the lost horse, but was proceeding to camp, where he would await my arrival.   Since we were already short of horses and had some three hundred miles to travel before we could reach the nearest settlements on the coast, aside from the value of the horse, I did not like the idea of losing his services.   I felt certain that he was somewhere in the forests in the vicinity of our last camp, and since it was already late in the afternoon, I decided on camping for the night on Mayer River and returning the following day to look for the lost animal.   Early the next morning I saddled my horse and, leaving the remaining two picketed, returned to our last camp.   As I emerged from the forest into the little open space in a bend of the stream where we had made our temporary stop, I caught sight of the object of my quest.   Immediately the horse saw me, he came trotting toward me with such an expression of joy depicted on his countenance as I have seldom seen in any of the dumb animals.   I was not long in returning with him to my camp of the night before, where, after hastily preparing and partaking of a breakfast of hot coffee, bread, and a grilled venison steak, I packed up and continued across Mayer River on my return to camp, where I arrived with all the horses and paraphernalia in good condition, much to the delight of Mr. Peterson, who had passed a rather uncomfortable night through being somewhat short of bedding.

We remained a few days at this camp completing our collections from the fauna and flora of the immediate vicinity.   Thus far I have said little concerning the nature and variety of the trees that in this region form the forests of the Andes.   And indeed there is little to be said, for on the eastern side of the main range they all belong to one species, the deciduous beech, *Fagus antarctica.* Clinging to the branches were great bunches of *Myzodendron punctatum,* a beautiful green parasite resembling very closely our common mistletoe. A number of species of edible fungi grew attached to the branches and trunks of the trees, including the handsome orange-colored *Cyttaria darwinii,* which grows in small spherical masses with convoluted surface. This forms a staple article of food with the Channel Indians of the west coast and Tierra del Fuego. With the exception of the beech, no plant grows in these forests that is worthy of being called a tree. Indeed, the next largest plant is a species of black currant, which, under exceptionally favorable conditions, occasionally grows to a height of some eight or ten feet. On the outskirts of the forest the trees are generally dwarfed and ill-shaped, but within they frequently attain no inconsiderable dimensions, reaching to a height of forty or fifty feet, with a diameter near the base of three, four, or even five feet, or more in rare instances. The heart-wood is of a reddish color, of fairly fine grain and capable of receiving a considerable degree of polish. About Sandy Point, Dawson Island, and at several stations in Tierra del Fuego, considerable quantities of this wood are being manufactured into rough lumber, though it is really better adapted for inside than for outside use. If properly handled, there is no reason why it should not make a fairly good lumber for cabinet purposes. When subjected to the combined effects of the atmosphere and moisture, it rapidly decays and soon becomes quite useless, so that the dead timber is always entirely ruined by decay. As fuel it burns readily and makes a quick fire, but the coals do not hold fire for any considerable length of time. Owing to this last property it should be especially valuable for the manufacture of matches, for once the flame is extinguished, the fire is out immediately.

Growing upon the ground, on fallen and decayed logs and about the trunks and branches of trees, are a considerable variety of mosses and Hepaticæ, while these plants, together with several species of ferns, are also common among rocky ledges and on the shores and in the waters of the smaller branches of streams, though not nearly so numerous or diversified as in forests along the channels of the west coast. On the lower outskirts of the forests there grows a considerable variety of small shrubs, one of the most abundant of which is a large shrubby member of the Compositæ, made conspicuous by its abundance of beautiful white blossoms.

About the middle of March we decided to set out on our return journey to the coast. During the month and a half spent within the Andes we had experienced most delightful weather. There had been numerous snow squalls on the mountains about us, but in the valley where we were encamped the weather had, as a rule, been all that could be desired. The thin ice that each morning covered the small water-holes, as well as the autumnal colors which the foliage of the forest was beginning to take on, bore no uncertain evidence as to the changing season. Moreover, since we already had a collection sufficient to try to the utmost the means at our command for its transportation to the coast, it was needless to prolong our stay in the Andes, however pleasant it had been. Packing our materials into as compact a form as possible, we started on our return journey. We had gone scarcely a mile up the winding course of the stream by which we had descended, when, while descending a short, steep slope, our cart turned completely upside down, breaking off both the shafts close up to the body. We were thus left in somewhat of a predicament, with a shaftless cart, five hundred miles from the nearest repair shop. However, we were fully prepared for just such emergencies, and with axe, brace, bit, monkey wrench, and a few other tools, in a couple of hours we had hewn and fitted a new pair of shafts from two of the forest saplings, which were in every way better than the old ones, and we were once again on our way. We travelled leisurely along, stopping at each of our old camps to take on such material as we had packed and left on our journey into the mountains. As we passed along beneath the high bluff, near the summit of which was the dinosaur limb I had found shortly after our arrival, I earnestly wished to add the fossil to our collections, but our limited means of conveyance forbade the gratification of the desire.

Passing along the edge of the last of the forest-covered slopes, we emerged from Mayer Basin through Shell Gap into the broad, level valley of the Rio Chico. It was with feelings of regret that we looked back upon the wooded slopes and snow-clad ranges of the Andes where we had passed so many pleasant and profitable days. We would fain have remained longer to enjoy the primitive condition and novelty of this region rich in nature's handiwork. But already the frequent falls of snow on the slopes about us, the ice-covered pools which greeted us on each successive morning, and the rich autumnal colors of the forests, reminded us of the approaching winter and that it was time we were on our way across that three hundred miles of uninhabited country which lay between us and the nearest settlements on the coast.

We returned by practically the same route as that by which we had come, descending the Rio Chico until we reached Sierra Ventana, where, instead of turning to the south across the low plain which separates the valley of this stream from that of the Rio Chalia, or Sheuen, we kept down the Rio Chico until some fifteen miles below the mouth of the Chalia, when we crossed over to the Rio Santa Cruz.

As we encamped for our last night on the Rio Chico, we were visited by a band of Tehuelche Indians. They were greatly interested in us and our equipage, were extremely anxious to learn from whence we had come and the nature of our journey. For the most part they spoke Spanish quite fluently, though we were somewhat deficient in the use of that language. They plied us with all sorts of questions relative to ourselves and our mission, and when we hesitated or interrogated them as to the meaning of some word which we did not fully understand, they would reply: "Usted no entende esta palabra? Esta palabra cristiana" (You do not understand this word? This is a Christian (Spanish) word), showing much amusement at our ignorance of what they supposed to be our own language. They were apparently quite unaware that there were more than two languages — Tehuelche and Spanish. However, they were extremely good-natured and jovial, as indeed I always found these Indians to be. I never observed any of that morose nature among them which is so characteristic of many tribes of our North American Indians.

Our trip down from the mountains was easily accomplished and was attended with only minor difficulties and delays, until we arrived at Las Salinas, near the head of tide water, on the Santa Cruz River. Here we were delayed for several days in crossing the river. Owing to the difficulty of crossing our cart and horses we availed ourselves of an advantageous opportunity for disposing of them, and had ourselves, our collections and collecting materials transported down the river to the village of Santa Cruz in one of the boats of the "Cross-Owen," an English schooner that happened at that time to be lying in the port.

After spending a few days in Santa Cruz, packing our collections preparatory to their final shipment to New York, we went on to Gallegos, where we arrived early in May, 1897, which place we had left on our trip into the interior in December, 1896. We had, therefore, been absent just five months, and during this time we had not only received no news from the outside world, but after leaving the Santa Cruz River on our trip north we met with no person, either Indian or white, until on our return journey we reached the settlements near the mouth of the Rio Chico. It had been one long, delightful journey over vast plains, along the base of high basaltic platforms, across swollen rivers, through the deep and winding, basalt-capped cañon of the River Chico, over the continental divide, and down the Pacific slope into the very midst of the virgin forests that flanked the mountains in the interior of the Andes. There was little of adventure, but much of novelty and interest, so that time passed only too rapidly.

CHAPTER IX.

Death of Governor Mayer; Voyage to Tierra del Fuego; Forests of Vil-
larino Bay; Beagle Channel; Ushuaia; Lapataia; Wreck of the
"Esmeralda"; St. John's; Giant Kelp; Return to Buenos Aires; Sail
for New York; Arrival at Princeton.

BEFORE our arrival at Gallegos we had learned of the death of our
friend, Governor Mayer. His death was universally regarded as a
loss, not only to the Territory of Santa Cruz, but to the Argentine Republic. On our arrival at Gallegos we received our first letters
from relatives and friends at home, whom we had left fourteen months
earlier. They brought news of the sad death, on the eighteenth of the
preceding November, of my youngest son, a little boy just entering
upon his fourth year, to whom both myself and his uncle, Mr. Peterson,
were naturally much attached by many tender remembrances.

At Gallegos we learned that there was no steamer south of us, but that
one was expected at any time from the north. After a tedious delay of
some two weeks the "Villarino" arrived on her way south. Since there
would be no opportunity for us to go north until her return, we decided
to make the trip through the Straits of Magellan and around Tierra del
Fuego, rather than to await in Gallegos her return from the south. We
left Gallegos on the twentieth of May and at about noon of the following
day anchored in the harbor at Sandy Point, where we remained until
the morning of the twenty-second, when the vessel again resumed her
journey through the Straits. At about ten o'clock on the following morning the weather forced us to come to anchor in a land-locked harbor,
known as Villarino Bay on the coast of a small island a little to the
southwest of Tierra del Fuego. We were forced to remain here for a
couple of days, until the weather improved sufficiently to make it possible to navigate the narrow and difficult channels through which we
were to pass, and which extend almost continuously along the south
coast of the island.

During our stop at Villarino Bay some of the boats were manned and sent ashore to procure a supply of fresh water, and I availed myself of this opportunity for going on shore. As we approached the point selected as a landing place at the head of the bay, we passed close by a beautiful, pure white albatross standing on the surface of a low ledge of rock that extended for several rods out into the waters of the bay. It was a truly noble-looking and beautiful bird, although its pure white plumage seemed strangely out of harmony with its surroundings.

Having safely landed, I noticed a number of small rounded mounds, or hillocks skirting the shore at only a little distance from the beach. On examination these proved to consist chiefly of shells. They were in fact genuine kitchen middens, or shell heaps, still in the process of formation. On the summit of several were to be seen remains of charred wood and shells and on one I found a discarded and worn-out basket made of rushes woven together, remains of a small and dilapidated wickiup and other articles, indicating that within the last few weeks at least it had been made the temporary residence of a small party of Channel Indians.

After a few moments spent in examining these shell heaps I turned my attention to the forests. The shore was fringed with a dense growth of dwarfed and procumbent beech bushes. Through, or rather over these, I had to force my way before gaining the depths of the real forest, which lay beyond. When, however, after considerable difficulty I succeeded, I found myself, although in south latitude 55°, surrounded by a vegetation so profuse and abundant in its growth as to suggest that I had been suddenly transported into the midst of some tropical jungle. The palms and tree-ferns of the tropics were replaced by the Winter's bark, *Drimys winteri*, and the deciduous and the evergreen beech, *Fagus antarctica* and *F. betuloides*. Here, at this high latitude, on the twenty-fourth of May, equivalent to that day of November in our latitude, humming birds were busily engaged sucking nectar from the delicately colored bossoms of the *Philesia buxifolia*, while parrots disported themselves in the branches of the trees above my head. The earth was covered over with a profusion of ferns, mosses, Hepaticæ and lichens. The delicate, lace-like patterns of the fronds of several species of *Hymenophyllum* were not the least striking of the wonderful assemblage of cryptogams with which the ground was carpeted. Into the pile of this carpet, formed of these most exquisite and delicately colored plants, at each successive step I sank to a depth of several inches. The brilliant green of the ferns and several species of mosses and Hepaticæ was mingled with golden-colored Hypnums, and the richly tinted whites, pinks, oranges and yellows of a multitudinous variety of lichens and fungi. Nor was this beautiful assemblage of colors and patterns limited to the surface of the ground; the trunks of the trees were likewise covered with them, while they hung in long festoons from the lower branches like beautiful draperies, frequently arranged with faultless taste. The trees were, generally speaking, of considerable dimensions, several feet in diameter, but rather low and with numerous large horizontal branches. So low and abundant were these branches in many places, that they offered a serious obstacle to my progress through the forests and compelled me to leave the ground and take to the branches, where for several hours I walked about, stepping from one limb to another, frequently many feet above the earth. While thus engaged with these novel experiences, and in the midst of such interesting surroundings, I was so completely shut in and protected by the dense foliage of the uppermost branches of the trees, that I was only occasionally reminded of the fierce storm which was raging without by the increased rumbling above my head, due to the particularly violent "williwaws" that at intervals descended from the snow-clad mountains in the rear and swept down through the forests and out over the turbulent waters of the bay. The delicate foliage kept perpetually fresh by frequent showers and the variety and harmony of colors displayed by the ferns, mosses, Hepaticæ and lichens, which grow in such profusion in the depths of these forests, are most pleasing in their effects. They enhance the other natural beauties of this region and give to the quiet depths of these sylvan retreats a peculiar attractiveness, contrasting strongly with the rugged cañons and serrated peaks of the higher Andes.

Since I had been assured by the ship's officer that I need be in no hurry and might take my own time for returning to the vessel, I passed most of the remainder of the day, quite alone, in the solitude of this truly wonderful and primitive forest, returning in the evening with my botany press, pockets and arms filled with botanical specimens, not a few of which proved to pertain to previously unknown species.

During the evening of the twenty-fourth of May the storm subsided, and as the morning of the twenty-fifth dawned, we were steaming leisurely through the western stretches of Beagle Channel, one of the most remarkable inland waterways of the world.   At times we would steam along for several miles through intricate narrows, never more than a few hundred yards in width, and with lofty mountains rising on either side precipitously from the water's edge.   Then, again, the channel would open out into a broad reach of water, across which, as we proceeded on our way, a considerable portion of the adjacent parts of Tierra del Fuego would be visible.   Descending the mountain slopes were to be seen numerous glaciers, some of these reaching quite to the water's edge. During the day we passed Mt. Darwin, the most conspicuous of the many prominent peaks along this coast.   The weather throughout the day was extremely changeable.   At one moment the sun would be shining brilliantly, with scarcely so much as a breeze to disturb the surface of the water.   In an incredibly short space of time these conditions would be completely altered, as the skies became overcast and a storm swept down from the adjacent mountains, bringing with it a blinding blast of sleet and rain, which, as it passed over the channel, would, for a short time, completely obscure our view of the surrounding shores and drive us to seek shelter in the ship's saloon.   Such conditions were, as a rule, quite temporary, and, in a short time, the sun would again shine forth, the clouds rapidly disappear, gradually disclosing to our view the slopes and summits of the adjacent mountains.

This being the twenty-fifth of May and a national holiday in Argentina, we were treated in the evening to a champagne dinner.   The wines were good, but no amount of clarets, sauternes or champagnes, whatever their quality, could overcome the strong flavor of garlic which pervaded everything of an edible nature with which the ship's tables were supplied.   Not only were the meats and vegetables seasoned with that offensive herb, but the pastry as well was flavored with it.

On the evening of the twenty-sixth of May we reached Lapataia, just within the western boundary of Argentine Tierra del Fuego.   Here we remained for only a few hours, when we proceeded some twelve miles farther on to Ushuaia, the capital of that portion of the island which belongs to Argentina.

Ushuaia may justly lay claim to being the southernmost town in the world.   It lies at the head of a small bay of the same name, and at the time of our visit there were perhaps twenty or thirty rude buildings and about one hundred inhabitants.   It was first settled by English missionaries, sent out about 1860 by the Church of England Missionary Society, to do mission work among the Yahgan Indians, at that time the most populous tribe inhabiting this coast and the adjoining islands. The work of these missionaries has been continued with little interruption throughout the past forty years, and with remarkable success, so that the Yahgans, who formerly numbered several hundred, have been so thoroughly civilized by the missionaries that, according to a statement made to me by Mr. Lawrence, who at the time of my visit was in charge of the mission at Ushuaia, they are now reduced in number to some thirty or forty souls. This depopulation has been due chiefly to consumption, propagated and fostered among them, it seems very probable, by the sudden introduction of clothing without, at the same time, instilling into their minds those hygienic principles essential to the proper use of such garments.

Beside being the capital of Argentine Tierra del Fuego, Ushuaia gains additional importance from being the site of a small sawmill and a factory for the canning of *Lithodes antarctica* and *Paralomis granulosa*, large crabs found in the waters of this region. We went ashore and spent the day looking about over the town and in a walk through the adjacent forest, which had been already somewhat devastated and shorn of its primitive nature by the axe of the woodsman, marks of which were to be seen on every hand. In the town we were surprised to see, in one of the gardens, a bed of pansies in full bloom, though it was now late in May, a season almost equivalent to the first of December in our latitude. Here, as elsewhere in these forests, the two species of beech, *Fagus antarctica* and *F. betuloides*, were the prevailing, and, except for an occasional example of *Drimys winteri*, the evergreen Winter's bark, the only real trees to be found.

After a day passed at Ushuaia the "Villarino" returned to Lapataia, which in a small way has been made a coaling station by the Argentine government. There was also a sawmill at this place, and we remained here for three days, recoaling and embarking a quantity of lumber and telegraph poles for the proposed telegraph line along the east coast of Patagonia. This much needed convenience, which could be constructed with very little difficulty, has as yet never succeeded in getting beyond the prospective stage. It furnishes one among many striking examples of the dilatory methods pursued in Spanish-American countries. If the same conditions were to arise at any time within territory belonging to the United States, as have existed for the past fifteen years in Argentine Patagonia, within sixty days, at the outside, a telegraph line would be built and in operation, bringing all the ports of this coast into immediate connection with the home government and the outside world. We passed a considerable portion of the time at Lapataia in the forests adjacent to the little land-locked bay, where we made considerable additions to our botanical collections.

After three days passed at this port we proceeded on our way eastward through Beagle Channel to Bridges Station, where we stopped just long enough to send a boat ashore with the mail and then continued our journey, until we arrived at a point on the south coast of Staaten Island, where a German bark, "the Esmeralda," had a few days before met with adverse winds and been driven on shore and wrecked. Our vessel was brought to and a boat lowered and despatched to the disabled vessel. The boat succeeded in reaching the bark, but found her deserted. Deciding that the crew had made their way by land or water to St. John's, a convict and life-saving station on the opposite side of the island, for which port we were also destined with supplies and recruits for the station, we were not long in proceeding on our way.

When we arrived at St. John's, we found that the crew of the wrecked vessel had made their way to that port, where they had been picked up a few days earlier by a passing steamer. We remained at St. John's for an entire day. The harbor is very small and is entered by a very narrow channel. I had at this place no opportunity for going ashore, which I should very much like to have done. The waters of the bay were exceptionally quiet and beautifully clear and transparent, so that it was easily possible, by looking over the vessel's side, to see for several fathoms beneath the surface. There was an abundance of kelp, *Macrocystis pyrifera*, growing about in the water, and the more isolated specimens of this wonderful plant formed most beautiful objects, as the broad, flat fronds, furnished with pyriform air-sacs, floated gracefully on the surface of the transparent water, beneath which the long, spiral, rope-like stems could be seen descending for several fathoms to their place of anchorage on the rocky bed beneath. This plant is one of the wonders of the vegetable kingdom. It frequently attains to a length of several hundred feet. It is very abundant everywhere about the coasts of these islands, where it has been of the greatest service to the navigator, since every submerged rock supports a growth of it, which acts as a buoy, warning the experienced mariner of the danger which lies hidden beneath the surface. The waters of the little bay at St. John's were also rich in animal life. A peculiar, almost transparent, umbrella-shaped jelly-fish, and certain large star-fishes were especially abundant. In the evening the bay seemed fairly alive with seals, which disported themselves in great numbers in the water about the sides of the vessel.

We left Staaten Island in the evening, and the following morning, in the teeth of a strong westerly breeze, we arrived at San Sebastian Bay, on the east coast of Tierra del Fuego. Here we stopped for a couple of hours to exchange mails before proceeding on our way to Gallegos, where we arrived on the fourth of June, after having completely circumnavigated Tierra del Fuego. At Gallegos we bade good-bye to our many friends, and on the following day, June fifth, went aboard the "Villarino," homeward bound. At Santa Cruz we took on the collections made during our trip into the interior, and, after the customary stops at the various ports of call along the coast, we arrived at Buenos Aires, June twelfth.

On enquiry at the Lamport & Holt offices in Buenos Aires we found that the "Maskelyne," one of that company's steamers, would be sailing in a few days for New York. We immediately engaged passage on this vessel, and the morning of June seventeenth found us on board and steaming slowly out of the "Boca" and down the River Plate to Montevideo, where, for three days, we were delayed, completing our cargo of dry hides and charqui for New York and Cuba. Leaving Montevideo on the twentieth of June, we proceeded directly to New York, stopping only at Castries in St. Lucia, a small island in the West Indies, which derives its chief importance through the advantages offered by its small but excellent harbor as a coaling station.

As we approached Cape St. Roque on our homeward voyage, we encountered the strong south equatorial current, which for several days increased our normal run for each twenty-four hours by from fifty to seventy-five miles. When we arrived opposite the mouth of the Amazon, although distant more than three hundred miles from shore, the waters of the ocean were still much affected both in temperature and color by that mighty river. We re-coaled at St. Lucia, and proceeded on our way to New York, where we arrived on the evening of July sixteenth, just too late to be received and disembark with our baggage. On the following morning our vessel was taken to her berth at Martin's Pier, and after most courteous treatment at the hands of the customs officers we were soon on our way to Princeton with our somewhat numerous packages of most unordinary baggage, most of which, over the railroad, we were compelled to send by express or freight.

CHAPTER X.

Start on second expedition; Sail on Grace Line steamer "Cacique" direct to Sandy Point; Arrive at Sandy Point; Encamped in the forests on the Rio de las Minas adjacent to Sandy Point; Coal Mines at Sandy Point; Otway Water; Start North; Visit Tehuelche village; North Gallegos; Santa Cruz; Start for the interior; Stuck fast in the Rio Chico; We leave the Rio Chico near the Andes and explore the country to the northward; Mt. Belgrano; Over the high plateaus; Spring Creek; Swan Lake; Discovery of Lake Pueyrredon; Arroyo Gio; Rio Blanco; White Lake; Evidences of extensive ice action in the region lying east of Lake Pueyrredon; Remarkable geological section in bluff south of Lake Pueyrredon; Abundance of both vertebrate and invertebrate fossils; Birth of the southern Andes; Caught in a blinding snow storm on the mountain; The valley lying east of Lake Gio; In the basalt cañons southeast of Lake Buenos Aires.

WHILE our work had been most successful, there remained much to be done, and I immediately set about preparing for a second expedition, on which I was able to embark on November ninth, 1897, after a stay of nearly four months at home. On the second expedition Mr. Peterson remained at Princeton, engaged in freeing from the matrix portions of the material already collected, and I took with me as assistant Mr. A. E. Colburn, a most kind and obliging young man from Washington, D. C., who in addition to having had considerable experience as a taxidermist, had previously been on two expeditions, one to Newfoundland and the other in Florida.

Profiting by the experiences of our first expedition, I took with me on the second a light two-and-three-quarter-inch mountain wagon with a good pair of double harness. Taking a Grace Line steamer, the "Cacique," we left New York on the morning of November ninth, bound direct for Sandy Point in the Straits of Magellan, which was to be our first stop on the outward voyage. During this long voyage of thirty-two days after leaving Sandy Hook, we sighted land but once, at Pernambuco, until our arrival at the eastern entrance to the Straits of Magellan. On December eleventh we came to anchor in the roadstead at Sandy Point and the following day disembarked with our cargo, taking quarters at a hotel in the town, until such time as we should be able to procure horses suitable for the prosecution of our work.

After a few days we succeeded in purchasing a pair of work-horses and with these we left Sandy Point and established a camp in the forests some four miles above the town on the Rio de las Minas (River of the Mines). At this camp we remained for nearly two weeks. Mr. Colburn found the woods about our camp excellent collecting ground for birds, while my own time was fully taken up with the flora of the region and in studying and collecting fossils from the extensive and excellent geological section afforded by the walls of the cañon through which the river flows. Several hundred feet of Tertiary deposits are exposed in the sides of this cañon, in which are represented a number of different and quite distinct horizons. For the most part, they are quite fossiliferous, and from these I secured a considerable series of invertebrate fossils. The beds are marine throughout, with the possible exception of certain coal-bearing horizons toward the middle and top of the series, which may possibly in part be of brackish or fresh-water origin. In the middle and upper series there are a number of veins of very pure lignite. These veins of lignite vary in thickness from a few inches to ten or twelve feet. A twelve-foot vein, which crops out on the left bank of the river about five miles above the town, was opened up and worked somewhat extensively about thirty years ago, when an attempt was made to make of Sandy Point a coaling station for steamers passing through the Straits. The attempt proved unsuccessful, however, and had to be abandoned on account of the low calorific properties of the lignite and the difficulties and loss sustained in attempting to coal steamers from open lighters in an unsheltered harbor like that of Sandy Point. Though finally abandoned, it was not until after a railway had been constructed from the port to the mines and a drift driven some three hundred yards into the latter. At the time of our visit the railway had long suffered from disuse and much of the original road bed was entirely washed away, while the locomotives had been dismantled and the other rolling stock was in a most dilapidated condition. An attempt, however, was then being made by a wealthy Chilian firm to reopen the mines and make of Sandy Point a profitable coaling station by substituting for this lignite briquettes manufactured from an admixture of lignite and English or Welsh coal. The work was in the hands of a competent engineer, and already the mines had been reopened and the drift driven still farther into the hill, while a diamond drill had been imported, with which to prospect the underlying strata in hopes of finding a better quality of coal, which would obviate the necessity for the manufacture of briquettes. At a depth of some six hundred feet drilling was suspended without meeting with any coal of the desired quality. I have never learned what became of the project for the manufacture of briquettes, but since the local manager of the company was strongly opposed to it, I presume it was abandoned. This lignite is very pure, and while not of a quality sufficiently good to commend it for sea-going vessels, it would doubtless serve very well for all domestic and light steam purposes. Its supply is well nigh inexhaustible.

During our stay at this camp I rode one day up the cañon, through the forests and out upon the bare, rounded summit of the mountain which lies between Otway Water and the Straits of Magellan. The top of the mountain presented a series of rounded hillocks separated by small marshes and swamps. The surface was covered over with many huge glacial bowlders and supported a growth of low subalpine plants. The timber line was several hundred feet below the summit, though I should not estimate the altitude at above two thousand feet.

Otway Water is the southernmost of that intricate series of inland waterways which enter the mainland from the Pacific and about the nature and origin of which I shall have more to say when I come to treat of the geography of Patagonia. It is about fifty miles in length, with a very narrow entrance, but expanding within into a broad pear-shaped body of water, lying on the north side of the Brunswick Peninsula, the southernmost point of the mainland of South America. Sandy Point is situated on the eastern shore of the isthmus which connects this peninsula with the mainland.

Having completed our work near Sandy Point, we started early in January on our trip to the north, going by way of Gallegos and by the same route as that by which I had travelled on my first visit to Sandy Point during the previous expedition. We travelled leisurely and stopped at various places along the route to collect birds, mammals and plants.

At Gallegos we purchased a second pair of work horses and lead-harness, and thereafter drove four horses to our wagon.

From Gallegos we visited some Indian villages on the upper course of the south fork of Coy River and distant about seventy-five miles. Our purpose in visiting these Indians was to secure a series of photographs and procure material illustrating their arts and industries sufficient for reconstructing a family group for the United States National Museum, that institution having given us a commission to do this work. The first village visited consisted of some eight or ten toldas and perhaps forty or fifty people, of which number not more than one half were of pure Tehuelche stock. Of the women only three were living with Tehuelche men. The remainder claimed as husbands either men of impure Indian extraction or nondescript Europeans. Among the latter I noticed that here, as elsewhere, those of French, Spanish, and Portuguese blood predominated. In families where both the parents were of pure Tehuelche stock, there were few children. In no instance do I remember having seen a family of pure Tehuelches in which there were more than three children, while one or two was the more usual number, and frequently there were none. On the other hand, in families of mixed blood the number of offspring appeared to be about normal.

The site of the village at which we encamped was directly on the bank of the stream in the midst of a wide open valley, offering no protection from the winds and where fuel was extremely scarce.

During the first day or two of our visit we were treated rather coolly and were unable to make much progress with our work. However, after a couple of days, by a judicious distribution of tobacco among the men, and raisins, sweet chocolate, ginger snaps and other similar articles with which we had supplied ourselves, among the women and children, we were able to overcome their disinclination toward us and to establish ourselves on terms of friendly intimacy with them. We soon experienced little trouble in securing most of the more ordinary articles such as rugs, bridles, saddles, bolas, etc. Such articles as were the property of adults could be acquired with little difficulty, while in most instances no amount of money would tempt them to part with the fur mantle, rattle, or other childish toy belonging to one of the children. For a time, through their timidity, we were unable to secure any satisfactory photographs. After a few days, however, I succeeded in getting one of the young women to

FIGURES 19 AND 20—SEE OTHER SIDE

19—Tehuelche cradle

20—Tehuelche woman prepared for a journey

stand for her photograph. From this I made a few prints and gave her. These had the desired effect, and thereafter, so far from being refused the privilege of photographing the different groups and individuals, I was fairly besieged by those wanting pictures taken of themselves.

So jealously did they guard every article connected with the life of a child that, for a time, I almost despaired of securing a collection which would in any way adequately illustrate the method of transporting, rearing and caring for the young child among these Indians. It was not that they regarded such objects as of any very great intrinsic value, nor that they were averse to our becoming the possessors of them. Indeed it was quite evident after a few days that, aided by our camera and confections, we had quite won the hearts of all the children and a considerable number of the ladies of the village, while, by a generous distribution of tobacco and cigarette papers, we had likewise gained the confidence of the men. We were especially anxious to get the cradle shown in Fig. 19. So important was this article in every family where there was an infant, that we felt our collection would in no wise be complete without an example of one of these most convenient and useful articles, together with its accoutrements. It not only serves as a receptacle for the helpless child when in the tolda, but when on the march with the child fastened inside and covered over with warm fur rugs, when the whole is securely lashed on to the horse behind the mother, as shown in Fig. 20, it forms a convenient conveyance for both child and cradle, while moving about from one place to another over the Patagonian plains. Our first attempts to secure this cradle resulted in an absolute failure, notwithstanding that we offered in Argentine currency one hundred and fifty dollars — about seventy American dollars, while its intrinsic value could not, at most, have been more than three or four dollars. Nor was our attempt to secure other cradles less unsuccessful. Fortunately, however, on the evening previous to the day we had set for our departure the father and mother of the child to whom the cradle belonged came to us with the pleasing information that, since their child had almost outgrown the need of the cradle, and as they were, for the present at least, not expecting the arrival of another occupant, they were willing to dispose of it and for a much less sum than that which we had already offered. We were not long in striking a mutually satisfactory bargain, and early the next morning were on our way down Coy River to North Gallegos, where we stopped long enough to pack up our materials thus far collected and leave them with our friend, Mr. Halliday, to be shipped by the first vessel which should call at that port.

From Gallegos we proceeded to Santa Cruz, and, crossing the Santa Cruz River some twenty-five miles above its mouth, started on a second trip into the interior in search of the Pyrotherium beds of the Ameghinos. For some two hundred miles our course lay up the River Chico, following practically the same route as that by which we had descended that stream on our first trip. In crossing this river at a rather high stage of water our team stalled in the middle of the stream, where we had to leave our wagon standing with the water running half way over the box, while we carried our equipment and supplies to the opposite shore on horseback, and later succeeded in drawing our wagon ashore by attaching picket ropes to the end of the tongue. We spent the succeeding day in overhauling and drying our outfit and supplies. A considerable portion of these had become thoroughly wet while in the river. We were not a little chagrined to find that most of our salt and a considerable portion of our sugar had been dissolved by the water, while there was scant satisfaction afforded by the reflection that, in bulk at least, the loss in salt and sugar had been fully recompensed by the expansion which had taken place in the contents of a case of evaporated apples that chanced to be stored near the bottom of the wagon. Our matches, however, caused us most concern. As for salt, we could still supply ourselves from the lagoons of the pampas. Of sugar we still had plenty, for we had started from Santa Cruz with a one-hundred-pound sack. Since neither of us smoked, we did not need many matches, and as we still had a few small boxes of dry ones, we decided to go on with these and dry such as had been wet, in hopes that some of these at least might still be serviceable. In this we were not unsuccessful, and we were, therefore, not without a sufficient supply of those useful articles.

We crossed the Rio Chico a few miles below the mouth of the Rio Belgrano, just where the wide valley, mentioned in the narrative of our first expedition, enters the river valley from the north and distant about forty miles from the eastern base of the Andes.

Leaving the river at this point, we laid our course almost due north over the level surface of the broad, open valley, which gradually but rapidly increased in elevation and assumed more the appearance of a high pampa or plain than that of a valley. On the west this valley was enclosed by a lofty, basalt-capped table land, which extended uninterruptedly for a distance of forty miles. At the southern end this basaltic platform had an elevation of one thousand feet or more above the valley, but at its northern extremity the difference in elevation was scarcely more than two hundred feet. On the east of the valley lay the great lava fields, which in this region cover most of the interior of the Patagonian plains. From certain patches of green in the vegetation, discernible at none too frequent intervals just below the basaltic platform on our west, we knew that water for ourselves and horses could be had, if we kept along the base of this escarpment. We, therefore, crossed the valley to a point just beneath the first of these "mananteals," believing not only that we should find an abundance of water for our purposes, but also that the surface of the valley at the foot of the cliff would afford us a convenient highway for our journey northward. In the first of these suppositions we were not disappointed, but we were quite mistaken as to the latter, for instead of the valley continuing as an unbroken and level surface to the foot of the bluff, as we had supposed, it was interrupted throughout the entire extent of the latter by a series of deep depressions, usually from one to two or three miles in diameter and frequently several hundred feet in depth. These depressions were quite similar to those occupied by the salt lagoons already mentioned as existing throughout the eastern plains region. They were usually separated from one another by a narrow isthmus extending from the level plain to the foot of the bluff. The bluffs surrounding these depressions were, for the most part, rather precipitous and not easy of descent and ascent with our wagon, so that we had usually to keep to the eastward of them as we proceeded on our journey. The origin of many of these depressions, or "sink holes," was a question to which I gave much attention while in Patagonia, but was never able to arrive at any very satisfactory conclusion. They will be more fully discussed when treating of the geography and geology of this region. There was an abundance of water at all the mananteals, though it was only to be reached by a long and laborious climb for several hundred feet up the steep incline, covered over with huge blocks of lava derived from the basaltic platform at the summit. The fauna and flora in and about the springs of these mananteals were quite varied and differed exceedingly from those of the surrounding country. They afforded most excellent collecting grounds, and at our first camp we remained for several days to collect botanical and zoölogical materials and rest and shoe our horses. The contact of the basalt, here from two to three hundred feet thick, with the underlying sedimentary rocks, determined the positions of the springs of water. About the margins of the depressions and over the lower of the débris-covered slopes of the escarpment were frequent exposures of sedimentary deposits belonging to the Santa Cruzian beds.

While encamped at this locality, I climbed one day to the summit of the basaltic platform and travelled for a considerable distance over its deeply fractured and cavernous surface. From my elevated position I secured a splendid view of the surrounding country. To the eastward, beyond the valley across which we had last driven, lay the great lava field through which, for two hundred miles, the Rio Chico had carved its deep and winding cañon. From my more elevated position I could look down upon and across its broken and uneven surface, which appeared not unlike that of a great sea of black water in a state of violent disturbance. Looking southward beyond the valley of the Rios Belgrano and Chico, there appeared, near the western border of this great interior lava field, a number of old volcanic peaks with summits already covered with snow. On the west was a wide strip of open country, characterized by high rolling divides separating the river valleys, over which were scattered numerous glacial hills, interspersed with small lakes and broad stretches of meadow lands. Such was the country that lay between the high basaltic outlier on which I stood and the yet higher and noble mountain ranges, whose snow-capped summits rose ever upward, until lost in the clouds on the western horizon. On the north, beyond the limits of the basalt field which lay at my feet, the landscape assumed the nature of an elevated plain, out of which, at a distance of some fifty miles, Mt. Belgrano raised its mighty mass in solitude, towering to a height of some three thousand feet or more above the surrounding plain. From the nature of the country to the northward it was evident that we should encounter little trouble for a considerable distance, while travelling in that direction. Turning to my more immediate surroundings, the surface of the lava was seen to be composed in places of highly vesicular, scoriaceous materials, occasionally approaching in character that of a true pumice. In other places it was quite hard and firm, while there were occasional nodules of a black obsidian with an even luster and conchoidal fracture. There was little vegetation and less of animal life. While the vegetation was, for the most part, limited to a number of lichens and a few scattered tufts of grass, at one place in the sides of a rather deep cavern I found a small fern growing in considerable profusion. By turning over a number of large detached pieces of lava I procured a number of beetles, crickets, spiders, a centipede and a scorpion, while small black lizards were not wanting.

The basaltic escarpments and the slope below were favorite haunts of the condor, and one morning, while strolling about among the latter, I came upon five of these splendid birds, and was somewhat surprised at their temerity. Walking up to within some forty feet of them without causing any alarm, I seated myself on a rock for a few moments, when I was struck with the remarkably fine subject they would make for the camera, and returned to camp for that instrument which I had left behind. When I returned to the place where I had left them, they had quit the locality, and I never succeeded in getting a similar view of these birds at such close range.

We continued our journey along the eastern base of this basaltic table, until arriving at its northern border. Beyond this there extended for many miles a wide, open plain, the level surface of which for a considerable distance was interrupted only by the shallow channels of a couple of dry water-courses, which unite and pass through a narrow defile in the lava fields to the eastward. Pursuing as nearly as possible a direct northerly course, after crossing the channels just mentioned, we ascended a considerable incline and gained the summit of the broad and level plain which lay beyond. As we continued our journey across the plateau-like table land, with Mt. Belgrano rising on our left at a distance of some fifteen or twenty miles, like a huge and solitary sentinel, from the surrounding plain, we gradually placed the lava fields, between which we had been travelling, in our rear, while in front and on our right there extended a broad, open country, which seemed, within the limits of our horizon, to be almost perfectly level.

After travelling for some twenty miles across this plain, we came suddenly upon the crest of a bluff, overlooking a rather deep but narrow valley, which, at a distance of about a mile to the northeast, opened into a broad, level valley about ten miles in width, through which there meandered a small stream, Spring Creek on the map. Descending into the smaller and tributary valley, we camped at the bottom alongside a beautiful spring of most excellent water, where there was an abundance of grass for our horses.   Here, as subsequently became our custom, now that we were travelling through an unknown country, we remained for a number of days, while I examined the geology and geography of the surrounding region and Mr. Colburn employed himself with the birds and mammals of the more immediate vicinity.

During our journey across the table-land, before arriving at the cañon, I had detected in the distance to the eastward what I thought to be extensive exposures of sedimentary rocks.   The following morning I proceeded on horseback to determine the extent and nature of the supposed sediments. Following along the crest of the bluff above the main valley, it was seen to have an average width of about ten miles.   Through this valley there ran a small stream, Spring Creek, at first with a considerable current. As I descended the valley, however, the stream became successively more sluggish, until it finally terminated in a small lake with a diameter of perhaps a mile.   An examination of the shores of this lake showed that it was subject to periodical overflows when, by melting snows or heavy rains, the volume of water became so much increased that it overflowed the lower border of the lake, where a rather deep channel had been cut through the upper part of the retaining embankment.   Below this lake the valley extended for several miles, with numerous springs along its northern border, until ending in a lake some three or four miles in diameter, almost entirely surrounded by a series of rugged bluffs, consisting of igneous materials, barren brown sandstones and dull red porphyries, the latter not unlike in general appearance those already mentioned as occurring at Port Desire on the Atlantic coast.   A careful and continued search among the sandstones was unrewarded by the discovery of any fossils, save a few uncharacteristic plant impressions.

The lake was, for the most part, quite shallow, but in its waters there grew a profusion of aquatic plants, while the dead shells of a species each of gastropod and bivalve Mollusca were literally piled in diminutive winrows along the beach, testifying to the abundance of animal life within its waters.   The plant and animal life so abundant in the lake made of it a favorite resort and feeding-ground for the various water fowl indigenous to this region, and at the time of this my first visit in early March, 1898, it was literally alive with these birds.   Duck and plover were there by thousands, while the American grebe and the flamingo were not wanting.   But the most striking feature of the avian fauna was the great number of swans. It is no exaggeration to say that there were on this lake during my first visit not less than a thousand of these most graceful birds.

In numbers the black-necked species greatly predominated, but those of a pure white color were not entirely wanting. Owing to the abundance of these birds at this locality I have named this Swan Lake, and it is so shown on the map.

After spending several hours about Swan Lake, I proceeded to an examination of the country to the eastward, the surface of which was found to be entirely occupied by the great lava-sheet of the central plains region. The lake had no outlet and it was quite evident that it had been formed by the damming of an ancient watercourse, which formerly had extended uninterruptedly from the present lake and valley above, through a deep cañon cut in the lava beds to the eastward. I traced the deep and narrow valley of this now desiccated stream for several miles to the lava fields, and, although immediately below the lake its channel is filled for a depth of one hundred feet, or more, with sedimentary materials, it was quite evident that at some previous time it had formed the channel of a continuous and not unimportant watercourse.

In some exposures along the south side of the valley lying south of Swan Lake I discovered a number of unimportant fossils, sufficient, however, to determine the beds to belong to the Santa Cruzian formation. At night I camped in a thicket of calafate bushes by the side of a spring some ten miles south of the lake, sleeping in my saddle-blanket and slicker and dining off the body of a red-breasted meadow lark grilled over a bed of glowing coals. Hardly had I unsaddled and picketed my horse when I was attracted by the peculiar and playful antics of a specimen of the little gray fox, *Canis azaræ*. Attracted by the presence of myself and horse, this beautiful animal had left his retreat and, prompted by curiosity no doubt, came boldly up to within a distance of some thirty or forty feet of me, where, in evident satisfaction with my companionship, he ran and frisked about in a manner quite like that of a favorite domestic dog, and much to my amusement. I permitted the sport to continue for several minutes, then, drawing my revolver from its scabbard, I despatched the beautiful animal and determined, as I had suspected, that he was a young male, though scarcely less than an adult in stature.

The locality chosen for my solitary resting place during the night proved also to have been in times past a favorite encampment for the Indians. Scattered about over the ground were numerous fragments of broken pottery, stone scrapers, drills and arrow points.

On the following day I examined a number of small exposures in the sedimentary deposits for fossils and returned to camp in the afternoon.

The following morning, with a fresh horse, I started to explore the country to the northward. Crossing the main valley of Spring Creek, I ascended a high ridge on the opposite side. The surface of this ridge, which at this place was some five miles in width, increased rapidly in elevation to the northward and was covered over by numerous small round hills, chiefly composed of fine sand and gravel, but with many large polished or angular blocks of granite, syenite, gneiss and other crystalline rocks of unusual size, frequently attaining to a weight of several tons. The whole aspect of the deposit was such as to suggest that the materials could have been transported to and deposited in their present resting place only by ice. Continuing my journey through these glacial hillocks, I came finally, at a distance of about five miles, to the crest of a high escarpment overlooking the deep, broad valley of Arroyo Gio, lying east of a small lake bearing the same name. Following westward along the crest of this escarpment, it was found to increase rapidly in elevation and from a particularly advantageous point I got my first view of the eastern extremity of Lake Pueyrredon, of which at that time I believed myself to be the discoverer, and which in a short paper I subsequently called Lake Princeton, in honor of Princeton University. It had, however, been discovered during the same season, but a few months earlier, by Señor von Platen, an engineer of the Argentine Boundary Commission. Though Lake Princeton has priority of publication, I gladly relinquish it in favor of Pueyrredon, the name given by the Argentine commission after a vessel in the navy of that country. To Señor von Platen belongs the credit for the discovery of this magnificent lake, with a length of fifty miles and an average breadth of from five to ten miles.

Descending to the bottom of the valley, which lay some two thousand feet below the crest of the bluff, I continued in a northwesterly direction, to examine what appeared to be a bad-land area, distant some ten miles from the foot of the bluff and in about the middle of the valley. These bad lands proved to be the bluffs of the Rio Blanco, a small stream flowing into White Lake. At the point where I reached the river, it flowed along the bottom of a deep and rugged cañon, the walls of which were composed entirely of glacial débris, consisting of a heterogeneous mass of silt, gravel and huge masses of rock, for the most part, crystalline and frequently weighing several tons each. The cañon had a depth of some three hundred feet or more, and its sides, as well as the stream at the bottom, were quite picturesque. The materials composing the bluffs were, from their origin and nature, especially interesting, but of course quite devoid of vertebrate or other fossils. On leaving camp in the morning I had intended returning the same day, but it was nearly sundown when I climbed the cañon wall and gained the level surface of the valley above. Out of consideration for my horse I stopped for the night where there was an abundance of grass and bushes, near a spring which issued from the side of the escarpment to the south of the valley, and at an elevation of about one thousand feet above its bottom. It was long after dark when I arrived at this place, and since I had neither taken the precaution to bring food with me on leaving camp in the morning, nor provided myself with meat during daylight, I was compelled to go to bed (?) supperless. Having picketed my horse on water and grass, I built a rousing fire by an adjacent bush, and about this passed the greater portion of a most pleasant night. There was an abundant supply of dry brush convenient, and from this I could with little difficulty replenish my stock of fuel. It was an ideal autumnal night, and, stretched on my saddle blanket before the comforting warmth of the glowing camp-fire, I gave myself up to a full enjoyment of the conditions surrounding me. It is wonderful how under such conditions one's mind in the end involuntarily assumes a reflective mood. Every trivial detail of the past life is recalled and friends and places long since forgotten are brought most vividly to the memory and always in most agreeable form. Then, at times, one passes insensibly from the reminiscent to the contemplative state, and a consideration of the past gives way to that of the present and future. Thus, quite undisturbed, one mentally surveys the present conditions and surroundings, passing on to the elaboration of most praiseworthy, but seldom realized, plans for the future. Those who have a true love for nature must at times find this affection so strong as to drive them beyond the limits of civilization to some retreat where, unmolested, they may study her in her true form and beyond the environmental influences of man. The night was far spent when, aroused from my reveries and wrapped in saddle blanket and slicker, I gave myself up to sleep.

The next morning I continued my journey to camp, stopping long enough to examine a thick layer of basalt which cropped out just below the top of the bluff and was overlaid by the glacial deposit already mentioned as occurring on the summit. On my return to camp I left the glacial hills and entered the broad valley of Spring Creek, just at a point where there were a number of fine springs. Here I decided to establish our next camp, until such time as I should be able to examine the country some thirty to forty miles to the westward in the vicinity of Lake Pueyrredon, where from a distance I had already observed that there were considerable exposures of what appeared to be sedimentary rocks.

On the morning of the following day we moved across the valley of Spring Creek, camping at the springs just mentioned. That same afternoon I started for Lake Pueyrredon, intending to return the following day. Holding a more westerly course than I had done on the last excursion, I descended into the valley some three miles east of where Rio Blanco enters it, after emerging from a cañon in the southern border. For some distance after entering the valley this river flows along over a shallow, rocky bed, before entering the deep and narrow cañon cut in the glacial drifts a few miles below. Crossing this stream I directed my course for a bare rocky elevation lying between White Lake and the eastern extremity of Lake Pueyrredon, which I reached late in the afternoon. Directly opposite this point a small river, which I have called the Rio Tarde, for want of a better appellation, after emerging from one of the most picturesque cañons in this region, turns abruptly to the westward and empties its waters into Lake Pueyrredon. The valley here has a width of perhaps not more than five miles, and is for the most part well watered and grassed. Unsaddling and caring for my horse, I set out to examine the nature of the rocks constituting the mountain mass lying between White Lake and Pueyrredon. I found these to consist principally of variously colored sandstones and porphyries, frequently of most brilliant colors, and among which certain masses of a peculiar green color predominated. The character of this rock was in every way similar to those pebbles described by Darwin as of a gall-green color and so abundant over certain regions in the shingle of the plains near the coast. There is little doubt that this was the source of many of the pebbles seen and described by Darwin. The rocks in reality form a considerable dome-like, mountainous hogback, which, with an altitude of perhaps fifteen hundred feet, is continued out into the middle of the valley at right angles to the general trend of the main range. Its surface is deeply scored and polished and these features, together with the glacial materials left stranded at an altitude of fully three thousand feet above the bed on either side of the main valley, bear unmistakable evidence that at some previous period the ice of a mighty glacier filled this great depression and overflowed on the surrounding hills, extending eastward over the plains for a distance of fully sixty miles, as I afterwards discovered by tracing the terminal moraine. The enormous erosion and deposition accomplished by this mass of ice is partially indicated by the several hundred feet of glacial material shown in the walls of the cañon of the Rio Blanco. These glacial materials I subsequently observed to extend eastward to a distance of from forty to fifty miles.

The very nature of the materials constituting the rocks of this mountain precluded the possibility of my finding fossils, but the remainder of the day was well spent in an examination of their lithological nature and in learning what I could of the geography of the region. Late in the evening I returned to my horse, where, in the shelter of a clump of bushes, I retired for the night.

The following morning, after an early and hastily prepared breakfast, I continued my journey, directing my course toward the southeastern extremity of Lake Pueyrredon. Hardly had I started, when, from a shallow depression in the surface of the valley in front of me, there appeared a small column of smoke rising into the atmosphere. This was interesting, for since leaving the settlements near the mouth of the Rio Chico, we had seen no one. Continuing in the direction from which the smoke was seen to rise, I soon came in sight of three men and a number of horses. On conversation with the men I found them to belong to the Chilian Boundary Commission, engaged in a preliminary survey, with a view to determining the boundary line between Chilian and Argentine Patagonia. From the first it was quite evident that there was mutual surprise at our meeting in such an unexpected time and place. After a few moments spent in conversation with these men, the only persons beside my companion that I saw during this my second trip into the interior, I bade them good-bye and continued my journey toward Lake Pueyrredon.

This lake lies in a deep valley, at an elevation of scarcely more than one hundred meters above the level of the sea. To the eastward this valley extends for many miles, connecting with that of White Lake and the basalt cañons still farther to the east. The divides separating Lake Pueyrredon from White Lake and Rio Blanco and the latter from the basalt cañons lying east of Lake Gio, are very low and consist chiefly, if not wholly, of glacial materials. The valley proper is continuous and forms one deep, broad and uninterrupted trench, extending eastward from Lake Pueyrredon for a distance of some fifty miles, where it is abruptly contracted and thence becomes a deep, narrow defile, continued for an unknown distance, with rugged enclosing walls, composed of igneous and sedimentary rocks of varying age and character. Immediately south of the eastern extremity of Lake Pueyrredon a rather precipitous bluff rises abruptly to an altitude of five thousand feet above the waters of the lake, as shown in Fig. 21. The rocks composing this are of Cretaceous and Tertiary age and of sedimentary origin, save some fifty feet of basalt which separates the uppermost Cretaceous from the lowermost Tertiary material. The different strata composing the bluff are, for the most part, well displayed in the face of the cliff, while, as I subsequently learned, there are extensive exposures of the uppermost beds in a region of bad lands at the summit.

I reached the eastern end of the lake at about nine o'clock in the morning, and, after riding a short distance along its shore, admiring the beautiful body of water as it lay spread out before me, its silvery surface unruffled by even the faintest zephyr, at a convenient place I unsaddled and picketed my horse, and, with tools and collecting bag, essayed to climb the bluff that towered above me and the lake. The task proved a greater undertaking than I had anticipated, but by perseverence and patience after several hours of hard climbing, interrupted by frequent stops, to examine and collect fossils from some particularly promising exposures, I arrived at the summit. Never have I seen a more delightful and pleasing view than that which greeted me. The morning, though calm, had been clouded and altogether uninviting. As the day advanced, however, the clouds vanished and the sun appeared, bathing the surrounding mountains with its autumnal warmth, while, as from a highly polished mirror, its rays were reflected from the surface of the lake that rested peacefully a mile beneath me. Nor was the splendid view afforded by my commanding position

21 – Lake Pueyrredon.

22 – A basalt cañon plains of Patagonia.

FIGURES 21 AND 22—SEE OTHER SIDE

the only or most substantial reward for my long and arduous climb to the summit. I had been repaid by the discovery of the most important continuous geological section I had yet seen in Patagonia. At the base were several hundred feet of Cretaceous materials, embracing several distinct horizons. Next followed some fifty feet of basalt, which in places exhibited a highly columnar structure. This in turn was overlaid by about one thousand feet of marine Tertiary deposits, rich in fossil remains, while above this came fifteen hundred feet of rock belonging to the Santa Cruzian formation, and at the extreme top there were from two to three hundred feet of marine beds belonging to the Cape Fairweather formation, already mentioned as first discovered at Cape Fairweather, distant some five hundred miles.

As I climbed my way up through the rocks belonging to the Santa Cruzian formation, I came frequently upon the bones and teeth of Nesodons, Typotheres, armored and unarmored edentates and other animals with which I had already become familiar. These were, as a rule, too fragmentary, however, to be of any value, except as determining the horizon of the beds, and, for the most part, they received only a passing glance. As I neared the summit, however, I came upon a place where a softer stratum of clay had weathered away more rapidly than the harder materials of the sandstones above and below. This had resulted in the formation of a level terrace several feet in breadth, on which I could walk comfortably, giving my attention to a survey of the surrounding rocks in quest of fossils, with no danger of losing my footing and tumbling headlong back down the precipitous cliff, as had been the case throughout a considerable portion of the distance through which I had climbed. I continued my walk but a short distance along this terrace when I came upon a splendid skull of Nesodon protruding from the face of the sandstone above me. This, on account of its size, prohibited my carrying it away. Continuing a little farther, I ascended to another and more extensive platform. Hardly had I reached the surface of this when I discovered an almost complete skeleton of a fossil bird of about the size of the blue heron, while close at hand lay another fossilized skeleton belonging to Diadiaphorus, an ungulate mammal, and near at hand lay several skulls and parts of skeletons of small carnivorous marsupials. I carefully took up and packed in my collecting bag the bird skeleton, a portion of the skeleton of Diadiaphorus and one or two of the marsupial skulls, and on looking about I was not long in discovering a number of others scarcely less interesting, which, for obvious reasons, I was compelled to leave behind.

As I continued my journey, arriving finally at the extreme summit of the bluff shown in Fig. 14, there appeared spread out before me a bad-land area several square miles in extent. The higher peaks of these bad lands I afterwards discovered to be capped with some three hundred feet of marine deposits, which, as mentioned above, belonged to the Cape Fairweather beds, according to Ortmann and Pilsbry, of Pliocene age. Through what enormous physical changes has the crust of the earth passed in this region within comparatively recent times! Here on the very slopes of the eastern range of the Andes, at an altitude of more than five thousand feet above the present sea level, are from two to three hundred feet of marine deposits of Pliocene age conformably overlying fifteen hundred feet of fresh-water, or perhaps in part, æolian deposits. And, moreover, the thickness and nature of these deposits, remarkably free from any coarse conglomerates, is such as would indicate that the birth of the Andes took place at the close of the Pliocene and that during the deposition of the Cape Fairweather beds a continuous, though shallow, sea prevailed over at least most of the region now occupied by the snow-capped peaks and ranges of the southern Andes, while during the Santa Cruzian period the same region was occupied by a broad, low and level land with numerous and, for the most part, small lakes, connected by sluggish streams and separated by broad marshes and uplands. At that time this region was inhabited by the birds and mammals whose remains are now found in such abundance in the sandstones and clays, in which they became imbedded, as the latter were laid down over the bottoms and flood-plains and along the shores of these prehistoric lakes, rivers and marshes. Where, then, it may be asked, was the land-mass from which were derived the materials composing these deposits? This and other questions will be considered when I come to discuss the geology of the region.

That at some past time, the five thousand feet of sedimentary deposits forming the bluff on which I stood had extended northward across the valley, was plainly evident from the nature of the surrounding country. That the valley had been formed by erosion was also evident. It was clear then that, since the close of the deposition of the Cape Fairweather beds, this entire valley, with an average breadth of ten miles and a maximum depth of a mile or more, had been scooped out by the combined action of water and ice, which latter agency had, as we have already seen, been not only destructive but constructive, having to the eastward left in the bottom of the valley a deposit of silt, covering the latter in some places, at least, to a depth of several hundred feet.

In the midst of such interesting surroundings time passed rapidly, and it had been long dark when, after many thrilling experiences attending my descent down the mountain, I arrived with my valuable and heavy load of fossils at the spot where I had unsaddled in the morning. Watering my horse and repicketing him on fresh grass, after one of the finest and most pleasantly and profitably spent days I experienced while in Patagonia, wrapped in saddle blanket and slicker, I retired for the night and, aided by the fatiguing experiences of the day, was soon fast asleep beneath the sheltering branches of a neighboring bush.

Encouraged by my success during the previous day and, notwithstanding that I had left camp with the expectation of only remaining over night, I resolved to see more of the bad-land area at the top of the cliff than had been possible in the limited time at my disposal. The morning broke cold, damp and disagreeable, but I was still hopeful that, as on the previous day, the weather might mend as the day advanced. In this respect, however, my hopes were not to be realized. I had noticed a small stream issuing from the bluffs and emptying into the lake on the south side at a point some four miles from its eastern extremity. From what I could see of the topography of the country I judged that it would be possible at that point for me to reach the summit with my horse, from whence I could explore the surrounding country without the necessity of returning to the valley in the evening. I was not mistaken in this surmise, but hardly had I covered one half of the distance to the summit when a blinding snow storm set in and continued throughout the remainder of the day and most of the night. The temperature was not sufficiently low to make the snow dry and crisp, so that it reached the surface in a half melted condition, remaining wherever it struck until quite melted. The wind blew fiercely and from every conceivable direction, driving the half melted snow into every nook and corner. About the middle of the afternoon, with every article of clothing thoroughly drenched, I arrived at a small beech forest near the head of the stream, along the right bank of which I had been travelling.   Here I unsaddled and picketed my horse in the grass on the outskirts of the wood, and in a sheltered place within I succeeded in starting a fire, though with considerable difficulty, about which I remained for the rest of the day and night, endeavoring to dry my shoes and other clothing.   Toward dawn the wind calmed and it grew colder, so that on the following morning the ground was covered with from three to four inches of snow, frozen sufficiently in most places to bear the weight of a horse.   The presence of the snow precluded the possibility of my spending the time to any advantage in a search for fossils in the adjoining bad lands.   Thinking, however, not to allow all my labor and hardships of the past twenty-four hours to count for nothing, I decided, rather than to retrace my steps and thus regain the valley, to strike out across the bad lands in the direction of camp, which I judged to be distant about sixty-five miles, hoping that in the course of my journey I might make some additional observations of interest concerning the geology or geography of the region.

I had proceeded only a few hundred yards, when a deer came walking out of a gully down the slope directly toward me.   The small lunch with which I had provided myself on leaving camp had long since been exhausted, and for some time I had been living on the bodies of red-breasted meadow larks and other birds grilled over an open fire.   This opportunity for securing a saddle of venison, though I could see that the animal was a doe, was most welcome, and I was not long in despatching her with my revolver.   What was my disappointment, however, to find, on going to dress the carcass, that it was not only a doe, but in milk.   A moment later I observed the fawn and a nearly grown yearling buck emerging from the same gulch from which the mother had but lately appeared.   Well aware of the superior quality of the flesh of a prime fawn as compared with that of an old doe in milk, I resolved to sacrifice one more life, and as the little creature in all innocence stood obliquely facing me, I sent a ball which I intended should strike just within and above the point of the shoulder and range diagonally through the body to the heart.   My aim was poor, however, and it took effect in the loin instead.   As I fired a second shot I did not notice that the third animal was steadily advancing so as to come in range with the fawn, and my second ball only entered the brain of the latter after it had come in contact with and carried away the entire symphysial region of the lower jaw of the former.   This necessi-

F<small>IGURES</small> 23 <small>AND</small> 24—S<small>EE OTHER SIDE</small>

tated my killing the third animal and the consequent extermination of the entire family, which I had in no way intended to do. With the saddle of the fawn and the hearts of all three I continued on my way.

I found the travelling through the half frozen snow and mud of the bad lands not only disagreeable, but extremely trying to my horse, and, after going a few miles, I came to the head of what appeared to be a considerable stream, leading in a southeasterly direction toward the valley of Lake Pueyrredon. I correctly judged this to be headwaters of the Rio Tarde and decided at once to follow its course and endeavor thereby to gain the valley running eastward from the lake, where I knew I should encounter much easier travelling than that which I had been experiencing in the upper country. For some miles my journey down this river was interfered with by no extraordinary difficulties. After a while, however, the valley narrowed and rapidly assumed all the characters of a mountain cañon, with walls towering on either side to a height of two thousand feet, or even more, while the river rushed madly through boiling rapids filled with giant bowlders, as shown in Figs. 23 and 24. I followed the course of the stream until further progress became absolutely impossible, when I was compelled to resort to a long, tedious and difficult climb to the summit of the bad lands above, where I arrived late in the evening. Here, with none too much grass for my tired and jaded horse, I passed the night without the kindly shelter of even so much as a bush to shield me from the piercing wind. I succeeded, however, in collecting sufficient dead roots and weeds to build a small camp fire, over which I grilled a venison steak, while in the hot ashes and coals remaining after the fire I placed the hearts which I had taken from the deer. These were nicely baked during the night and served me for breakfast the following morning. With saddle blanket and clothing both wet, I must confess to having passed a somewhat uncomfortable night.

By the following day the snow had disappeared to such an extent that I was able to learn considerable concerning the geology of the region. I passed most of the day in an examination of the lower Tertiary and underlying Cretaceous deposits, passing the night in the valley beneath and returning to camp the following day. I was extremely anxious to remain longer in this interesting region. Since, however, I had left my companion with the avowed intention of returning the next day and had now been absent a week, I feared he might be uneasy concerning

me.   Moreover, considering the jaded condition of my horse and the
state of the weather, it seemed advisable to return to camp, where I arrived late in the afternoon, somewhat to the relief of Mr. Colburn.   During the previous evening Mr. Colburn had enjoyed a visit from Dr.
Moreno who, with a party of six men and some sixty horses, had camped
with him for the night, while *en route* from Lake Argentino to Chubut in
the interests of the Argentine Boundary Commission.   I had made the acquaintance of Dr. Moreno at Santa Cruz and greatly regretted not being
present at the time of his visit.   Of all South Americans he doubtless has
the most exact knowledge of the geography of that continent and much
of it has the advantage of having been gained at first hand.

I had seen enough of the country lying about Lake Pueyrredon to convince me, not only that there was little hope of finding in that region the
Pyrotherium beds, which were the real objects of our search, but that the
altitude of the Tertiary beds in that vicinity was too great to permit
of their being advantageously worked during the present season.
I, therefore, resolved to move on northward.   The following morning
found us on our way north through the sand hills strewn with bowlders
toward the eastern end of the broad valley lying east of Lake Pueyrredon.
By bearing a little to the eastward we were able to skirt the basalt ledge
which outcrops at some distance below the crest of the bluff overlooking
the valley and to descend with little difficulty to the level surface of the
latter.   Once at the bottom, we travelled down the valley until reaching a
point where its width, not including the terraces on either side, is much
reduced and the wide marsh, which prevails for several miles above,
becomes a narrow stream, flowing between rather high bluffs formed of
glacial materials.   At this point we stopped for a day or two, in order to
examine the country to the eastward.

The bluffs of glacial material just mentioned as occurring on opposite
sides of the valley a little below camp, were clearly remnants of a terminal
moraine that had at some previous time completely dammed the entire
valley.   Some eight or ten miles below this the valley was again
obstructed by another moraine, which had not been so completely removed
as the first one.   Above this second moraine was a small lake some
two or three miles in length and perhaps half a mile in width.   Into this
flowed the stream above mentioned.   Below the lake the valley contracted
rapidly, becoming a deep and narrow basalt cañon, seldom more than a

hundred yards in width and with quite perpendicular walls, as shown in Fig. 22. The bed of the cañon was perfectly dry and whatever drainage, under ordinary conditions, takes place below the lake is subterranean in its nature, so perfect is the dam formed by the last terminal moraine. At various places in the walls of the cañon below the lake, the basalts were extremely porphyritic, and at many places there were outcrops of red porphyries similar to those already mentioned as occurring at Port Desire on the Atlantic coast. At other places were outcrops of the barren Cretaceous sandstones, while in deep channels cut in the eroded surface of the latter were to be seen deposits belonging to the Patagonian and Santa Cruzian formations.

After two or three days passed in this camp we moved across the valley and, by a rather circuitous route, gained the eastern extremity of a high basaltic platform, which extends eastward upon the plain from the foothills of the Andes lying to the north of Lake Gio. We camped for the night at some springs issuing from the base of these basalts, and the following morning, having skirted their eastern extremity, we continued northward over the plain, which, to all appearances, offered a perfectly practical highway for an almost indefinite distance. We were not long in discovering our error, however, for we had travelled little more than two hours in an east-by-north direction, when we encountered such a labyrinth of deep and almost inaccessible cañons as rendered further progress with our vehicle exceedingly difficult, if not well-nigh impossible. Descending into one of the shallower of these cañons to reconnoitre, I found near its head a small spring of most excellent water in a well-sheltered locality, with an abundance of wood for camp purposes and a plentiful supply of grass for our horses. At this spring, which was easily accessible, I decided to establish a permanent camp and explore the surrounding country.

This camp was located near the source of one of the numerous small cañons emptying into those marked Basalt Cañons on the map and lying east of Lake Gio and southeast of Lake Buenos Aires. At only a short distance below our tent this small cañon entered a narrow, picturesque and rocky defile, some views of which are shown in Figs. 9 and 10. With the exception of an occasional small spring, the bed of the cañon is dry. These conditions continue for a distance of one and a half miles, in which distance it receives several no less picturesque tributaries. It then enters the main basalt cañon mentioned above as continuing below the lake formed by the last terminal moraine crossing the valley which extends east of Lake Pueyrredon.   The bottom of this cañon is likewise destitute of any running water, but is remarkably picturesque.   It extends in an easterly direction for a distance of some three or four miles, where it enters the principal one of that intricate series of cañons, by which the surface of this country is dissected.   This cañon, unlike the others we have been noticing, is traversed by a considerable stream of beautiful, clear water well stocked with fish.   Its walls rise to a height of over 2,000 feet and the rocks composing them are made up of a rather complicated series of igneous and sedimentary materials, the age and nature of which, for the most part, I was quite unable to determine.   Figs. 7, 9, 10, 22, 25–28, 30 and 38 are scenes from this interesting region.

The point where the two cañons meet (Fig. 30) had in times past been a favorite Indian camping ground, as was plainly evidenced by the presence in great numbers of fragments of broken pottery, stone arrow-heads and other implements, as well as the broken and charred bones of fishes, birds and mammals.

During the first week of our stay at this camp I searched faithfully the surrounding country, within riding distance, for evidences of the Pyrotherium beds and fauna, but without any success whatever.   Then, taking with me an extra pack-horse, I explored the country to the eastward for fifty to seventy-five miles and northward as far as the eastern end of Lake Buenos Aires, likewise without success in so far as my work related to the discovery of the Pyrotherium beds and fauna.   In other respects, however, I was quite successful and secured considerable information relative to the geology and geography of the surrounding region, as well as important palæontological materials.   While thus engaged, it was my custom to take with me on each of my excursions, in addition to my saddle-horse, a pack-horse with tarpaulin, extra pair of blankets, and provisions sufficient for a couple of weeks, consisting of bread, rice, split peas, a small bag of pepper and salt mixed, tea, coffee and sugar.   For meat I depended on my revolver and the game birds and mammals of the country through which I was travelling, and my supply was seldom, if ever, deficient.

During my first excursion I passed some ten days in exploring the country to the eastward.   I first followed down the main cañon for per-

FIGURES 25 AND 26—SEE OTHER SIDE

haps fifteen or twenty miles, crossing and recrossing the stream as it zigzagged back and forth from one side to another of the narrow valley. For a time I experienced little difficulty in following the bed of the cañon. The stream had considerable current, and flowed over an almost continuous bed of coarse gravel consisting, for the most part, of crystalline rocks evidently derived from the bed of shingle which covered the adjacent plains. After a time, however, the current of the stream was arrested. It gradually became more sluggish and, owing to the soft and miry nature of the bottom, it was crossed with constantly increasing difficulty and danger, and finally resulted in the formation of an impenetrable swamp, through which the waters moved slowly on in deep, quiet stretches, with a current scarcely sufficient to indicate the direction of flow.

This swamp supported a luxuriant growth of rushes and extended on either side to the very base of the cañon walls, compelling me to leave the valley and take to the bluffs above, where, from the excessively rugged nature of the country, the travelling was extremely difficult. For more than a week I spent the time exploring that intricate labyrinth of chasms, with which the surface of this country has been so deeply dissected. The bottoms of the deepest of these chasms could scarcely have been less than three thousand feet below the level of the higher of the surrounding tablelands. Most of them were not only destitute of running streams, but it was quite evident that these conditions had continued for an indefinite period. In some of the larger cañons still occupied by flowing streams it was plainly evident that the farther from their sources one proceeded, the less important the streams became as erosive agents, while in the very midst of this deeply dissected region and where the depth of the cañons was greatest, the destructive work ceased entirely and they were absolutely engaged in constructive work, rapidly silting up their own valleys. From their very nature it was evident that these cañons had not been carved out within the recent geological cycle and by the small and insignificant streams, when any at all exist, that now occupy their beds, but that, for the most part, they were the remains of a previous drainage system that existed prior to the time which witnessed the great upheaval in the region now occupied by the Andes. This upheaval had resulted in an elevation sufficient to cause the accumulation of great bodies of snow and ice over that region, giving rise to the formation of glaciers that pushed eastward over the adjacent plains and temporarily filled the great valleys of erosion which had previously been carved in the surface of the latter, and permanently dammed the courses of the streams by the deposition of moraines, left at successive stages during the recession of the ice.

The walls of these cañons consist of a rather complicated series of igneous and crystalline rocks. The region bears evidence of having formed a land-mass during late Cretaceous and Tertiary times. In the walls of the present cañons were to be seen numerous ancient valleys cut deep into the surface of the older rocks, their troughs now filled with materials belonging to the Cretaceous or to the Patagonian and Santa Cruzian Tertiaries. Climbing to the summit of a particularly lofty, though isolated and limited table, I observed that to the east, north and south the same topographical conditions prevailed, and I was somewhat unwillingly forced to the conclusion that, during middle Tertiary times, this region existed either as a great island, or a narrow extension consisting of a chain of islands, belonging to a former great continental land-mass and surrounded by a shallow sea, over the bottom of which were deposited the Patagonian beds. In the later Santa Cruzian epoch this region became more elevated and the island, or islands, appeared as a low mountain range above the surface of a broad, level plain. It was then that erosion began and continued somewhat interruptedly, until it has produced the present deeply dissected condition of the country under consideration.

During my lonely wanderings through these deep and tortuous cañons I naturally met with many interesting things and some novel, though not always pleasant, experiences. Sometimes it would be the rescuing of myself and horses from the mire of a not very inviting swamp, through which we had essayed to pass to the bluff which lay at only a few hundred feet beyond, rather than submit to a circuitous journey of several miles. At other times, and all too frequently, the ascent or descent of the cañon walls was not without its perils and imparted a certain spirit of adventure to the work. This acted as a tonic and gave even a greater zest to its accomplishment. On one occasion, when I had camped for the night at the bottom of a particularly deep and narrow gorge where, however, there was a small stream and a plentiful supply of grass, I was startled from a sound slumber by the sudden stampeding of my horses. One of these I had left loose, while picketing the other, as had been my custom. It was with some alarm that I heard the two animals go dashing

27 – A basalt cañon, plains of Patagonia.

28 – Basalts of the Patagonian plains.

Figures 27 and 28—See other side

madly up the valley, after the bush to which the one was picketed had been torn from the ground. It required but a moment to thrust back the tarpaulin, slip on my boots and start in pursuit. The night was exceedingly dark, however, and in the depths of the cañon I could scarcely detect the nearest object. From the sudden manner in which they had become frightened, I knew that it had been caused by the approach of some wild animal, most likely a mountain lion, since these animals were very abundant in that region. As I hastened up the narrow defile in pursuit of the fleeing horses, I knew, by the sound of their iron shoes as they struck against the rocky bottom and reëchoed ever more faintly from the sides of the cañon walls, that they were rapidly distancing me. Still I kept doggedly on, for the thought of being thus left afoot was not a comforting one, and, aside from the loss of the horses, which, considering our remote position, would have proved a serious inconvenience, in addition it would involve the necessity of my traversing on foot the seventy-five miles which lay between me and our permanent camp. Owing to the darkness I had to rely more upon my ears than my eyes. The situation, however, presented one comforting phase. The inaccessible nature of the cañon precluded the possibility of their escaping from it, at least for several miles. Continuing my course, as I rounded a sharp turn, I again caught the sound of their footsteps. Stopping for a moment to listen, although owing to the darkness I could see nothing of them, from the nature and direction from which the sound came, I knew that they had ceased running and like myself had stopped to reconnoitre. I also knew that in their thoroughly frightened condition nothing but extreme caution on my part would enable me to approach them, without causing a second and perhaps more disastrous stampede. By careful manœuvring, however, I succeeded in making them aware of my presence, after which, much to the evident delight of both the parties concerned, and which they on their part showed by a succession of friendly whinnies, I had no difficulty whatever in catching and returning with them to my temporary camp.

The weather conditions throughout most of this trip in these cañons were all that could be desired. It was my custom to arise with the break of day, and, after a hastily prepared breakfast and a cup of coffee, set out about my work of examining the rocks of the surrounding cañons, making such notes and taking such photographs as seemed to be desirable. Sometimes I would be engaged in collecting sufficient fossils to determine the age of certain sedimentary rocks, which, like massive wedges, appeared filling deep trenches, that at some previous time had been carved in the surface of the underlying harder rocks, which now formed the greater part of the cañon walls.   It was my custom, as a rule, to take with me, while thus engaged, my horses and complete equipage, in order to avoid, as much as possible, the necessity of having to double back over the same territory.   As midday drew near, I would select some spot where grass and water were plentiful and turn my horses out to graze, while, beneath the shelter of a bush or ledge of rocks, I would indulge myself in a noonday nap, or, disinclined to sleep, would take the physical rest which the nature of my work demanded, while mentally engaged in contemplating the peculiar beauty, grandeur and solitude of my surround-ings.   On one particularly fine day, while engaged in this most pleasing and fascinating occupation, I was much interested in the peculiar action of a number of condors.   Since the day was warmed by a brilliant sun, not too often seen in Patagonia, I had thrown myself down upon my slicker in the comforting shade of a large calafate bush.   As I lay at ease, carefully scanning the cañon walls and the blue sky above me, my attention was almost immediately attracted by some half a dozen of those noble birds perched on the very summit of as many basaltic, needle-like pinnacles that rose perpendicularly from the opposite side of the cañon to a height of little less than two thousand feet.   With extended wings they were engaging in a series of very slow but rhythmical gyrations, turning slowly round and round without ever closing their wings or leaving their perch. So high above me were they that I could but just distinguish the white of their wings.   For fully an hour I watched their peculiar motions.   Evi-dently it was as calm at their altitude as in the cañon, and the day being an unusually warm one for Patagonia I attributed their actions to the heat, which to them had apparently become somewhat oppressive.

In the early part of the narrative of the first expedition I have remarked upon the peculiar manner adopted by the condor in attacking the carcass of an animal which has but lately died.   During my travels in this region I had an opportunity of verifying my former observations.   While travel-ling along one of the sides of the higher plateaus of this region I came upon a full grown guanaco that had become mired in a spring and was unable to extricate itself.   In its struggles it had fallen on one side, with its head resting on the bank.   In this position it lay, when by chance I discovered it. Although still alive, it was already surrounded by a number of carranchas and some five or six condors. One of the latter, as I approached, was observed to have taken a position on the body of the animal. It had already made a hole through the walls of the abdominal cavity, and was then busily engaged in tearing out piecemeal the intestines of the helpless but still living guanaco. Upon my near approach the birds withdrew, and I soon discovered that, as usual, the carranchas had been the first to detect the misfortune which had befallen the beast, for already the eye had been torn from that side of the head which was uppermost, though the tongue was yet intact. To send a ball from my revolver into the brain of the helpless animal, and thus end its sufferings, was a mission of mercy which I was not long in fulfilling.

While nowhere in the plains region of Patagonia had we seen the Chilian deer, *Cariacus chilensis*, yet I was not greatly surprised to encounter it here in a region which, though destitute of forests and distant from fifty to one hundred and twenty-five miles from the Andes, had all the characteristics of a rugged mountainous region, when one descended from the narrow, flat-topped tablelands to the bottoms of the cañons. I not only met with deer on various occasions in these cañons, but on returning to camp after this my first protracted journey in this region, as I was travelling up the chasm in which we had pitched our tent, I came suddenly upon a band of three at a distance of hardly more than half a mile from camp. Since we had thought of remaining where we were for the winter, this seemed an excellent opportunity for providing an ample supply of jerked venison, which is far superior to the flesh of the guanaco. As matters turned out later, it was perhaps unfortunate that my revolver hung so handily at my side, for hardly had the thought taken possession of me, when their three dead bodies lay on the talus-covered slope. Leaving the scene of slaughter, I went on to camp, returning a little later with Mr. Colburn, when, with his assistance, after the process of evisceration, we conveyed the carcasses to camp, where the flesh, skins and skeletons were properly cared for. In the hope of securing some of the Felidæ that were known to inhabit this region, we had poisoned the viscera, as had been our custom on other occasions. This proved disastrous to "Nig," our dog, who, while perfectly good-natured and absolutely worthless, had nevertheless the faculty of doing just those things he was not wanted to do. We had taken particular pains to see that he did not accompany us when we went for the deer, and since he seldom went far from camp, we felt little further concern.   Hardly had we seated ourselves for our evening meal, however, when we were attracted by his peculiar antics in front of our tent.   On emerging from the latter a glance at the dog showed only too plainly that he had been visiting down the cañon and had partaken of a portion of those choice morsels which we had prepared for other visitors.   Indeed, he was already suffering such violent spasms that Mr. Colburn begged me to shoot him and thus end his misery.   I was only too glad to comply with the request, since it involved only the consummation of an act which would, I fear, have been executed on any one of several previous occasions, but for a knowledge of the affection which had come to exist between Mr. Colburn and the dog.

On the day following my return to camp I started with fresh horses to explore the country to the northeast.   Here I found extensive and important deposits belonging to the Santa Cruzian formation, rich in fossil remains.   These deposits lie to the westward of the area of igneous and crystalline rocks, and between the latter and the Andes.   After several days spent in this region without discovering any trace of the Pyrotherium beds or fauna, I returned to camp and again set out to examine the region to the north and northwest, and lying between our camp and Lake Buenos Aires.   For several days during this trip my work was much interfered with by the great quantity of smoke from the forest fires in the Andes, lighted, as I afterwards learned, by members of the Argentine and Chilian Boundary Commissions.   So dense was this smoke that for several days I was unable to see in any direction for a greater distance than a mile or two, and it was quite impossible for me to distinguish promising localities and exposures from a distance, or to determine anything definite relating to the geography of the country.   On account of this interference with my work in the region about Lake Buenos Aires, I did not spend as much time as I should like to have done in the country lying immediately south of the lake.   From Lake Buenos Aires I returned to some important outcrops in the Santa Cruzian beds lying to the south ward between the base of the Andes and the basalt region of the interior. In these exposures I spent a part of two days collecting fossils and studying the geology of the region.   During the last night passed in these bad lands there was a considerable fall of snow.   This caused me to decide on returning to camp, which I did on the following day, April twenty-ninth.

CHAPTER XI.

Start on our return to the coast: Stricken with rheumatism: Receive our first news of the Spanish–American War, July 26th, 1898: Arrival at Santa Cruz: Mr. Colburn decides to return home and awaits steamer at Santa Cruz: I go to Gallegos on horseback: Caught in a blizzard: Across the ice-covered pampas: Reach the estancia of Mr. Halliday at North Gallegos: Cross the river to the Port of Gallegos; My rheumatism improves slowly because of insufficient comforts and I decide to return home: Embark for Sandy Point in the "Joseph F. Lovart" of the New York Pilot Fleet: An uneventful voyage of forty-seven days from Sandy Point to New York in the steamship "Maori": Arrive home much benefited by my long voyage through the Tropics: Arrange for my third and last trip to Patagonia.

FOR more than a month I had been faithfully engaged in exploring the country between and to the eastward of Lakes Pueyrredon and Buenos Aires in hopes of finding representatives of the Pyrotherium beds and their fauna. In the accomplishment of this I had not only been signally unsuccessful, but my search had been so thorough that I felt reasonably sure that they were not represented, except possibly by one or more small and insignificant exposures, anywhere, within a reasonable distance. During all this time I had passed but three nights in camp under the shelter of a tent. My work for more than two months, in fact ever since we had entered the practically unknown country north of the Rio Chico, had been continued from early morn until late at night, with irregular meals, and I generally passed the night with no other shelter than that afforded by a chance bush, and often with only my saddle blanket and slicker for a bed. As a matter of fact, through want of funds my entire work in Patagonia was carried on under most discouraging circumstances and with quite inadequate equipment, considering the magnitude of the task I had set myself to accomplish. With winter setting in and the chief purpose of this, my second visit to the interior, still unaccomplished, it was with rather depressed spirits that, on the twenty-ninth of April, I made my way through the snow, across cañons and over pampas, toward camp and my companion. Of provisions we had an ample supply to enable us to pass the winter, if we chose to do so. But from the elevation of most of the country it seemed clear that we should be able to accomplish little in winter. To the north of us lay a cañon, which with a wagon could be crossed only with the greatest difficulty. To the east was a region deeply dissected with a bewildering labyrinth of abysmal cañons. Under the conditions it seemed best to return at once to the coast by way of the Rio Chico, where I could profitably pass the winter by supplementing our collections from the Patagonian and Santa Cruzian beds, while Mr. Colburn could find employment collecting the water fowl of the region, to which we had hitherto devoted little attention. This was the plan I had mapped out as I rode along on my homeward journey, and which, after my arrival in camp, I unfolded to Mr. Colburn, and in the wisdom of which he concurred. The following day, April thirtieth, was passed in preparing for our retreat to the coast. Although unsuccessful in our main purpose, it must not be inferred that we had accomplished nothing. Mr. Colburn had secured an excellent series of the skins and skeletons of recent birds and mammals, while I, in addition to the knowledge gained of the geology and geography of the region, had made important collections of recent plants and invertebrates and had added considerably to our collections of vertebrate and invertebrate fossils, so that when ready to start on our return journey, our wagon was loaded almost to its full capacity.

On the evening of April thirtieth we retired with everything in readiness for an early start the following morning. When I arose the next morning my left knee was considerably swollen and quite painful, while the right was somewhat affected as was also my left arm. I regarded it, however, as only a slight attack of rheumatism, from which I had long been a sufferer, and which I believed would soon pass away without any serious results. At an early hour we were on our way southward. The wind blew raw and cold out of the southwest and chilled me through and through as I drove over the high pampa above our late camp in a direction almost facing the wind. It was late in the evening when we reached the stream in the valley lying east of Lake Gio. Here we camped for the night, and after applying a porous plaster, the only remedy at hand, to each of my knees, which were now both much swollen and quite painful, I was not long in seeking the comfort of my bed. I passed a feverish night, and the following morning the pain caused by the rheumatism in my knees and elbow was so intense, that it was only with the greatest difficulty I succeeded in harnessing the horses and gaining my seat on the wagon. My left arm was well-nigh helpless and had it not been for a desire to cross the high divide that lay between us and the valley of the Rio Chico before it should become covered with a fall of snow, I should have remained where we were until my condition improved. All day during the second of May I drove slowly southward, suffering the greatest pain as the wagon jolted along over the hummocky and uneven surface of the ground. At night we camped at a small spring on the divide between Spring Creek and the basalt cañons. The following morning my condition was so much worse that I decided to diverge somewhat from our course, in order to reach a number of springs near a fine meadow, which I knew lay a short distance to the west of Swan Lake. We reached these springs just before noon on the third day of May, and, after Mr. Colburn had pitched the tent and made down my bed for me, I retired in a most miserable and suffering condition. Both my arms and legs were now badly swollen, while my knees and elbows were especially painful, and, in addition, I was suffering with an intense fever. Notwithstanding my condition, however, I fully expected when I took to my bed, that I should be able to continue our journey within a few days. Such was not to be the case, for the rheumatism, together with the accompanying fever, rapidly increased in severity and soon spread to my hands, feet, neck and left hip. For six weeks I was absolutely helpless and unable to shift myself in bed or attend to my most trivial wants. Never has an invalid received more conscientious care than did I at the hands of Mr. Colburn. His care and patience were most exemplary, although I fear they were bestowed upon a somewhat unworthy person. If at times, as I fear too frequently happened, I appeared unnecessarily harsh or cross, I trust he will ascribe it to the great mental and physical pain with which I was afflicted, for looking back from this somewhat distant perspective I know and fully realize that to his tender care I am indebted for the privilege I now enjoy of making this humble, though grateful, acknowledgment of his kind attention.

So far from being able to proceed on our journey within a few days, days grew into weeks and weeks into months, and not until the twenty-ninth of June and midwinter had arrived, did I recover sufficiently to walk with a crutch, the product of Mr. Colburn's mechanical skill.   On that day, with the ground covered to the depth of a foot or more with snow, we left the camp where all unwillingly we had remained so long and started on our journey to Gallegos, distant some five hundred miles.   A view of our camp, taken on the morning of our departure, is shown in the photograph reproduced in Fig. 29.   I think my readers will agree with me after an inspection of this photograph, that the conditions were not particularly encouraging, when we were finally able to resume our journey. Day after day we dragged slowly along through fields of snow and ice, shovelling away the snow each night over an area sufficient to accommodate our beds.   We were frequently hard pressed to find grass sufficient for our horses, and the poor animals often suffered through a lack of sufficient food.   Before reaching the Rio Chico we were much disturbed lest, on account of the ice in the river, we should be unable to cross it.   Our fears were quite unwarranted, however, for on our arrival we found it so solidly frozen over, that we had only to drive across on the ice.

Our progress through the snow and ice was slow, Fig. 31 shows the nature of the country after two weeks' travel, and it was the twenty-sixth of July when we arrived at the settlements near the mouth of the Rio Chico and received our first news of the Spanish-American war.   The first report, given us by a Frenchman, was to the effect that Russia, Spain, France and Germany were at war with the United States and England.   This is cited as an example of the truth of the familiar adage that a story never loses anything during its travels.

When we arrived at the Santa Cruz River, we decided to leave our wagon and outfit and continue on horseback to Gallegos.   After crossing the river and reaching the Port of Santa Cruz, since a steamer was expected there within a few days bound for Sandy Point and Mr. Colburn had decided that he had had quite enough of Patagonia and would return home at the first opportunity, it seemed best that he should await the arrival of the expected steamer at Santa Cruz, while I proceeded on horseback to Gallegos.   Bidding each other good-bye, I started on the afternoon of the second day after our arrival at Santa Cruz for Gallegos, distant by land some one hundred and twenty-five miles. It was raining when I started, as it had been continually for the past three or four days.   Hardly had I gained the summit of the pampa,

29—OUR WINTER CAMP SOUTHEAST OF LAKE BUENOS AIRES.

30—CAÑON OF ARROYO GIO.

FIGURES 29 AND 30—SEE OTHER SIDE

which at a little distance west of the town rises to an elevation of some four hundred feet, when the inclemency of the weather increased and the rain descended in torrents, thoroughly drenching myself and my already jaded horses quite to the skin. On account of the melting snow and continued rains the surface of the level pampa had assumed all the characteristics of a quagmire. Instead of presenting, as in its normal state, a dry and almost barren surface, it appeared as a shallow swamp, almost limitless in expanse, through the deep mud and water of which my horses travelled with the greatest difficulty. At a distance of some thirty miles there was a sheep farmer's residence, where I had intended, on setting out, to pass the night. As the afternoon advanced, the fury of the storm increased, and toward evening the temperature fell and the rain changed to snow. This, driven by the terrific winds, only increased my discomforts and added to the darkness of the night which closed about me while yet several miles from my destination. I held patiently on my course, however, and about ten o'clock in the evening reached the estancia. Fortunately the occupants had not yet gone to bed, and, aroused by the barking of the dogs, two of the men came out and were only too glad to care for my horses, which, considering my benumbed and crippled condition, was indeed most welcome assistance. Within I found such comfort about the cheerful fire, as compared with the squalor of the best accommodations we were able to procure in Santa Cruz, made me almost glad I had braved the rigors of the storm. The good farmer's wife was not long in preparing for me an excellent meal, after partaking of which I retired for the night, thoroughly wearied with my long and tiresome ride.

Long before my arrival at the estancia the storm had assumed all the appearance of a genuine blizzard. This continued to rage throughout the night and the following day. Before morning the temperature had fallen to zero or below, and the snow was driven by the winds and piled in great drifts, completely covering the bushes and filling the smaller gulches, while the waters which had accumulated in broad, shallow lakes on the surface of the pampas were transformed into sheets of ice. So long as the blizzard continued, I could only remain as a welcome guest of the people at the estancia.

The morning of the second day after my arrival broke cold, clear and calm, and, notwithstanding the entreaties of my friends, through my great desire to reach Gallegos and receive the mail which I knew must be awaiting me there, I was forced to decline their further hospitality and proceed on my journey.   The high pampas were one broad sheet of glistening snow and ice, while the cañons were exceedingly treacherous from being choked with drifted snow.   In my crippled condition I could walk only with the greatest difficulty, and progress on foot was not only painful but exceedingly tedious.   To mount or dismount from my horse required great effort and was attended with considerable pain.   On account of the ice and the unshod condition of my horses, travelling on horseback, especially in my crippled and inactive state, was not without its dangers.   Frequently within the distance of a single mile my horse would receive several falls, and it was not an uncommon sight to see all five of the poor animals floundering on the ice at the same time.   From experience I soon learned that prudence demanded that, wherever a particularly smooth sheet of ice had to be crossed, I dismount and either lead or drive the horses in front of me.   So difficult was the travelling over the ninety-five miles of country that lay between me and Gallegos, that it was only by husbanding the strength of my horses in every possible way that I succeeded in reaching the estancia of Mr. Halliday at North Gallegos, late on the evening of the fifth day, with only one horse, having abandoned the others at various stages of my journey when, through fatigue, they became unable to travel farther.

It is needless to say that I was hospitably received and well cared for at the home of Mr. Halliday, whose hospitality, as well as that of his wife and family, are proverbial throughout Patagonia.   I learned from Mr. Halliday that a considerable quantity of mail was awaiting me at Killik Aike, which one of his sons volunteered to go for that same evening, a very thoughtful and highly appreciated kindness on his part.

I remained at Mr. Halliday's a few days and then crossed the river to the Port of Gallegos, where I found more mail awaiting me.   Although I had been greatly delayed in reaching Gallegos, I arrived ahead of the steamer for which Mr. Colburn had waited in Santa Cruz, and I had therefore the pleasure of seeing him on his arrival when the vessel called at Gallegos and of bidding him a second good-bye as he reëmbarked for his return to Sandy Point.

At Gallegos I received such medical assistance as was possible, and, securing accommodations at the best hotel in the place, I tried to recover from my crippled condition.   It was August and late winter, but notwithstanding the season, the inclement nature of the weather and my condition, I was quite unable to procure what I most needed — a warm and comfortable room. Indeed, not only was there no fire anywhere in the hotel, save the kitchen, but the building itself partook more of the nature of a dilapidated country barn than of a habitation for human beings. It was cold, damp and dirty, and nothing like as comfortable as a good tent. For such miserable accommodations I paid five dollars per day. I was too much crippled to walk about with any comfort, and, dress as warmly as I might, sitting about was a most uncomfortable occupation, so that in order to keep myself reasonably warm, I was compelled to take to my bed throughout a greater portion of each day, arising only for my meals.

As was to be expected, I did not improve very rapidly under such conditions. Moreover, the additional expense incurred by my illness was rapidly depleting the at no time very plethoric funds at my disposal, so that I was constantly harassed both in mind and body. As the weeks dragged wearily by I made little progress toward recovery. Late one night Captain Wilson, an American, who owned the "Estrella" (formerly the "Joseph F. Lovart," of the New York Pilot Fleet), a small sailing vessel which ran between Gallegos and Sandy Point, came to see me, as was his custom when in port. He was evidently not pleased with the progress I was making, and since, as he informed me, a Grace Line steamer was expected to arrive within a few days at Sandy Point on her way to New York, he was not long in persuading me to return with him to that port, where there was every reason to believe he would arrive a little before the steamer, and by which I could return to New York. This decision brought me to a painful realization of the depleted state of my funds, for after paying my hotel bill and a number of other smaller accounts, I found that I had only a few dollars remaining. However, my reputation for honesty was beyond reproach, and I had no hesitancy in applying to my landlord for a loan of fifty dollars until my return. As I had expected, this request was most cheerfully complied with. Captain Wilson assisted me in packing such articles as I decided to take with me, and, early the following morning, had myself and trunk put aboard the "Estrella." We left Gallegos with the early morning tide on September nineteenth, arriving at Sandy Point on the twenty-first, after a short and uneventful voyage. Here we learned that the "Maori," the steamer for New York, had not yet arrived, but was expected daily. On applying to Messrs. Braun & Blanchard, the Grace Line agents in Sandy Point, to engage my passage and enquire as to whether my personal check on the local Princeton Bank would be accepted in lieu of the cash for such passage, I was very agreeably informed that it most assuredly would, so that I had no further cause for worry on that score. On the afternoon of the twenty-second of September the "Maori" came in and dropped anchor in the roadstead, and late that afternoon I went aboard, since we were to make an early start the following morning.

Captain Eiley of the "Maori" had assured me that his vessel, although a chartered boat, was a "fourteen-knotter," and in every way superior to the regular Grace Line boats. I subsequently learned that, while his vessel may have been theoretically a fourteen-knotter, she could not in reality make seven. We left Sandy Point early on the morning of September twenty-third, and, after many delays, arrived at New York on November ninth, after a tedious voyage of forty-seven days. Our delays were due to our running aground in the shoal waters off Madonna Point, south of the River Plate, recoaling at Montevideo, where I spent the day very pleasantly with Colonel Swalm, the American consul, who had been acquainted with my father; a two days' delay off the coast of Brazil on account of disabled machinery; putting into Barbadoes through running short of coal; a delay of four days at St. Lucia, where we recoaled and the boilers were repaired, although they again broke down two days before our arrival and we were compelled to enter New York with but two boilers carrying steam.

Notwithstanding all these delays, which, under other conditions might have proved irritating, aside from the constant quarrelling between the captain and officers and the well-nigh intolerable provisions with which the table was supplied, I thoroughly enjoyed the long voyage through the Tropics. Moreover, it was just the tonic I needed, and of ever so many times more value to me than would have been untold quantities of iodide of potassium, salicylic acid, and other standard remedies for acute rheumatism, so that while I reached home still somewhat crippled from my severe attack, I was thoroughly refreshed in body, mind, and spirit, and immediately set about preparing for my third and final trip to Patagonia, without, however, applying to my friends for any additional funds.

CHAPTER XII.

Start on third expedition: Reach Sandy Point: Go by steamer to San Julian: Revisit Lake Pueyrredon: Discovery of new and highly fossiliferous Cretaceous deposits: Deep depression in plains east of Mt. Belgrano: A snow storm: Revisit Mayer Basin: Return to Santa Cruz: Make important collections of invertebrate fossils at mouth of Santa Cruz River: Return to San Julian: Gallegos: The wreck of the Villarino: Buenos Aires: Up the Paraguay River to Asuncion: Fossil bones at Entre Rios: Asuncion: The interior of Paraguay: Return to Buenos Aires: Rio de Janeiro: New York: Different effects exerted on the minds of Darwin and Hudson by the plains of Patagonia: Conclusion.

ON my third trip I was again accompanied by Mr. Peterson, and Mr. Barnum Brown, of the American Museum of Natural History in New York, also went with us in the interests of that institution. We sailed from New York on the morning of December ninth, 1898, just one month after my return. Our steamer was the "Capac" of the Grace Line, bound direct for Sandy Point, where we arrived on January tenth, 1899. From Sandy Point I went by steamer to San Julian, while Messrs. Peterson and Brown went overland to Santa Cruz by way of Gallegos. Between Gallegos and Santa Cruz they gathered the various horses I had left at intervals along the road on my trip during the previous winter and brought them to the latter place, where, as by previous arrangement, we all met at a date that had been mutually fixed and agreed upon. At San Julian I made a small collection of invertebrate fossils from the marine beds that outcrop at Oven Point, and with a saddle-horse and pack-mule and saddle set out to join my companions at the crossing of the Santa Cruz River.

From Santa Cruz we started for Lake Pueyrredon, following the same route as that by which I had travelled on my last expedition. On this last trip I had especially provided myself with extra saddle and pack-animals, and, as there were two persons beside myself in the party, there

was no reason why I should not be spared the irksome duties of a teamster and devote my entire time as we passed along to a study of the surrounding country.   Taking advantage of this opportunity, I carefully
examined the materials constituting the bluffs of the Rio Chico, determined the altitude of the basaltic platform and the upper limits of the
shingle at many different localities, beside making many other observations relating to the geology and geography of the region, which will be
of especial interest when I come to discuss the geologic and geographic
history of this district.

Our purpose in again visiting the country about Lake Pueyrredon was
to explore the region more thoroughly than I had been able to do during
my first visit, in the hope that we might find somewhere in that vicinity
exposures of the Pyrotherium beds, which, from the publications of Dr.
Florentino Ameghino, I had been led to believe might be quite possible.   We spent several days on the summit of the mountain lying south
of the lake in the midst of frequent and blinding snow storms and with
altogether most inclement weather.   Although failing in our principal
purpose, we secured a considerable collection of vertebrates and invertebrates from the Santa Cruzian, Cape Fairweather and Patagonian beds.
I was also successful in discovering a number of new and highly fossiliferous horizons in the Cretaceous at the mouth of the cañon of the Rio
Tarde, and at a second locality near White Lake, some ten miles farther
east.   From these I secured some thirty-six new species of invertebrates,
according to the determinations of Dr. Stanton, in whose hands they were
placed for examination and description.   The results of Dr. Stanton's
investigations form Part I. of Vol. IV. of the Reports of the Expeditions.

After some two weeks spent in the vicinity of Lake Pueyrredon, we
decided to return to the coast.   Messrs. Peterson and Brown, with the
wagon and outfit, returned by the route by which we had come, while I
parted company with them a short distance east of Lake Pueyrredon, and
with pack-mule and saddle-horse started off to the south to explore the country lying between the lake and the headwaters of the Rios Belgrano and
Chico.   On the day previous to our separating, while ascending one of the
lower benches of the bluff that rises above the valley extending east of the
lake, I observed a mountain lion that had been frightened from its place of
concealment and went galloping up the bluff and across the narrow plain at
the top.   As the country was an open one and I was mounted on a good

horse, this seemed an excellent opportunity and I was not slow in giving chase. Although the animal had several hundred yards the start of me, I rapidly gained on him, and when he reached the head of a small cañon at the opposite side of the narrow table I was not more than one hundred yards in the rear. On reaching the point where he had disappeared over the crest of the bluff I halted for a moment to reconnoitre. I knew the inability of this animal, like all the others of his tribe, to maintain any considerable speed for a long distance, and that when once beyond my sight he would seek refuge in concealment rather than flight. Over the slopes and bottom of the shallow cañon there was a considerable growth of scattered brush. By carefully scanning the ground about these I soon discovered the object of my search stretched at full length upon the ground. To despatch him with a rifle ball was the work of but a moment and required neither skill nor courage. I preserved both skin and skeleton, and, much to my surprise, they have been considered by Dr. C. Hart Merriam as belonging to a new subspecies.

Almost every traveller in Patagonia has remarked upon the naturally timid and cowardly nature of the puma. So far from a general disposition to attack man these animals are, as a rule, exceedingly timid, and examples are not at all uncommon where, when brought to bay, they have sought the shelter of a bush, and, without offering any real resistance, allowed the hunter to despatch them with his sheath knife, or by knocking them in the head with his bolas. Such timidity is not, however, universally characteristic of these animals, which are among the most abundant and by far the largest and most powerful of the Patagonian Carnivora. A notable exception to the rule, which came to the writer's knowledge, may be mentioned in this connection, since the facts connected with it are supported by unimpeachable authority. The case referred to is that of Señor Theodoro Arneberg, Chief Engineer in charge of the work of the Southern Division of the Argentine Boundary Commission. While engaged in his work in the vicinity of Lake Viedma in the autumn of 1898, in walking one day through a tangled mass of brush and tall grass, he came suddenly and unexpectedly upon a puma lying in concealment. The animal not only made no attempt to escape, but, instantly and without warning, attacked the intruder in the most savage manner. Springing upon him with its full force, it hurled him to the ground, although Mr. Arneberg is a large and powerful man, and the lion seizing him by the lower jaw, succeeded in breaking out several teeth and otherwise mutilating its then comparatively helpless victim, before one of his companions could rush up and despatch the thoroughly angered brute, which, after it had been killed, was found to be a very old male.

I had not travelled far on the morning after parting company with my companions to the eastward of Lake Pueyrredon, when it started raining. As the hours passed, the rain increased, so that along the foothills the bottoms of the usually dry cañons were soon occupied by swift-flowing streams, while the exposures over the slopes were made slippery by the softened clay. These conditions sufficed to make travelling through the foothills especially difficult, and, after a time, I was compelled to descend to the plains, which I reached just at the eastern extremity of Crystal Lake, a small but most beautiful body of water. Having reached the plain, I steered my course for Mt. Belgrano. The surface of the pampa was almost level and travel comparatively easy, until I came to a deep depression lying a little to the eastward of the mountain. I have else-where spoken of the prevalence of such depressions throughout the plains of Patagonia, but in depth this one surpassed any with which I met during all my wanderings in that country. My aneroid gave it a maximum depth of nine hundred feet beneath the nearly level surface of the plain, which extended southward from the west end of the depression. From crest to crest of the bluffs on the north and south its width was perhaps three miles, while its easterly and westerly dimensions were considerably greater, and the bluffs to the eastward, though still high, were much lower than those farther to the west.

The rain that had set in in the morning continued to increase through-out the day and, with the falling temperature toward evening, it changed to snow. Late at night I camped on the almost barren plain which extends southwestward from the foot of Mt. Belgrano. Here I passed the night without the shelter of a tent or even so much as the pro-tection of a bush. When I awoke the following morning, I could feel the heavy weight of the six or eight inches of snow that covered alike the surrounding plain and my tarpaulin. On thrusting my head from beneath the latter the sight which greeted me was, to say the least, any-thing but cheerful. The storm, so far from abating, had increased in fury and presented all the aspects of a full-fledged Wyoming blizzard. The temperature had fallen considerably during the night and a bitterly cold wind swept down the slopes of the Andes and over the level surface of the pampa. This was accompanied by a heavy fall of snow, which was driven by the terrific winds in every conceivable direction. The blinding snowstorm and constantly changing winds were calculated to have a most bewildering effect upon the traveller forced to continue his journey across the almost level surface of the plain. Such was the nature of the storm, that the most conspicuous landmark was not discernible at a distance of more than a few rods, while so variable was the direction of the wind, that it afforded an uncertain guide as to the course one should pursue. Confronted with such conditions, since, to use a military term, my present position was quite untenable, I was not long in saddling and packing my horses and renewing my journey across the plain in an attempt to reach a basaltic area which I knew lay some twenty-five miles to the southward, and where I was in hopes of finding shelter in some cave along the borders of the basalt and fuel sufficient to make a cup of coffee, for I had had no refreshment of any kind since the morning of the previous day. I knew that the prevailing direction of the wind was from the west and that if I kept it constantly on my right my course would lie in as nearly a southerly direction as it would be possible for me to hold under the conditions. I reached the ledge of basalt about two o'clock in the afternoon and succeeded in finding such shelter as I had anticipated. In a cavern in the basalt I passed a not very uncomfortable night and the succeeding day broke clear and cold, so that I was able to continue on my journey without further difficulty.

I directed my course toward some high hills lying about the head waters of the Rio Belgrano. The succeeding day was spent in these hills, though with little success, since the ground was still covered with snow. From here I continued southward across the north fork of the Rio Chico and on into Mayer Basin, which we had discovered on our first expedition some two years before.

Shortly after entering Mayer Basin, I had another one of those experiences which serve to illustrate the extreme temerity of the deer of this region, and which is undoubtedly due to the very limited contact which they have had with man. For some hours one afternoon I had been engaged collecting invertebrate fossils on the rather precipitous face of a considerable cliff facing the broad valley of Mayer Basin. Having secured such fossils as I desired, I descended to a convenient spot near the base and was busily engaged in properly labelling and packing my treasures, when my attention was attracted by the peculiar actions of my saddle-horse and pack-mule picketed in the valley close at hand. On looking about to discover the cause of their unusual display of interest, I discovered a fine buck approaching at a distance of some two hundred yards. From the manner in which he held his course directly toward me, it was evident that he had been attracted by my presence on the face of the cliff, and, prompted by curiosity, was intent on discovering the nature of an animal of such unusual appearance and habits. He came slowly and steadily toward me, until arriving almost at the very foot of the cliff and only a few yards distant. I was in need of a supply of fresh meat, and a shot from my revolver served to replenish my stock of that much-needed article.

I passed several days in Mayer Basin amidst scenes that had grown familiar during my previous visit. During my last night passed in the Basin, through a bit of carelessness I met with an accident that might have resulted far more seriously than it did. Early in the morning I had picketed my horse where both grass and water were plentiful. Then saddling the mule, I started out and passed the day in the surrounding hills. Returning late in the evening, thoroughly tired and with just time enough to prepare a little food before it became quite dark, I noticed the horse still picketed where I had left him in the morning, and remembering the excellent quality and inexhaustible supply of grass and water to which he had access, I decided that for once I would forego the usual custom of giving him further attention, believing that he would not suffer where he was until morning. I, therefore, unsaddled and hobbled the mule, got a bite to eat, and retired for the night. What was my surprise and chagrin, on awaking the following morning, to find that both mule and horse were gone. A little examination showed that the picket pin had been pulled during the night. I knew that as soon as they found they were at liberty, they had headed for San Julian, their old home on the coast, and distant some three hundred miles. With several hours the start, it was a source of little consolation to me to know that the mule was hobbled, for so accustomed had he become to those implements that they seemed an aid rather than an incumbrance to him. Discouraging as the situation was, it was not entirely hopeless, and I was not long in setting out at the best speed which I could reasonably hope to maintain for any considerable time, and at a distance of some twelve miles I succeeded in overtaking them. Returning I packed and saddled, and started at once on my journey down the Rio Chico to the Santa Cruz.

I arrived at the Santa Cruz some two weeks ahead of Messrs. Peterson and Brown, whom I passed on the way. Crossing over to the south side of the Santa Cruz River, pending their arrival, I passed the time in collecting invertebrate fossils from the locality in the Patagonian beds, at the mouth of the river made classic by Darwin. My work here was not only interesting, but very successful, and resulted in the discovery of several new species.

After the arrival of Messrs. Peterson and Brown I turned over to the latter such of my outfit as he desired to use in continuing his work, and he and Mr. Peterson went on south to work in the Santa Cruz beds along the coast, while I recrossed the Santa Cruz River and returned to San Julian. Here and at Darwin Station, some miles farther south, I spent two weeks collecting both vertebrates and invertebrates from the Patagonian and Cape Fairweather beds.

After a couple of weeks passed in the vicinity of San Julian the "Primero de Mayo," a small steamer belonging to the Argentine Transport Service, arrived from Buenos Aires, and on her I took passage for Gallegos. At Gallegos I stopped while the "Primero de Mayo" completed her voyage around Tierra del Fuego. By the time of her return I had completed all my arrangements for leaving Patagonia and with her I took passage for Buenos Aires, leaving Mr. Peterson to continue the work of collecting fossils in the Santa Cruz beds along the coast and to return by the next steamer arriving at Sandy Point and bound for New York.

We stopped at the usual ports of call between Gallegos and Buenos Aires, including Camerones. Here on a ledge of red porphyry, similar to that I have already mentioned as occurring at Port Desire and various localities throughout the interior, I saw the remains of the wreck of the "Villarino," the good ship that on our first expedition had safely carried us from Buenos Aires throughout the entire extent of the Patagonian coast and around Tierra del Fuego. Through sheer carelessness on the part of her officers, who, instead of attending to their duties while entering the harbor, had been attending the christening of an infant, born aboard the ship, the vessel had been driven at full speed hard on the sunken reef, literally ripping her bottom out, through more than half her length. Luckily she held fast to the reef, and since the weather was calm, no difficulty was experienced in disembarking the passengers and crew, and thus preventing any loss of life.

We arrived in Buenos Aires on the fifth of June. I remained in the city for a few days, renewing old acquaintances and forming new ones, after which I started on an extended trip up the river, going as far as Asuncion, the capital of Paraguay, passing and stopping at many interesting places on the way. The River Paraguay above its confluence with the Parana is still a noble stream, as it flows quietly along through broad savannahs or stretches of forest land, with banks elevated usually but a few feet above the water's level, as shown in Fig. 32.

At Entre Rios we stopped for a number of hours and, together with two of my fellow passengers, I took a drive through the city and visited an old monastery, where I was much interested in a considerable collection of fossils and other objects of natural history, which had evidently been picked up and brought together solely through a spirit of curiosity. Most of these bore no labels setting forth even so much as the locality from which they had been obtained. I was chiefly interested in the remains of some fossil saurians which, from the nature of the matrix in which they were imbedded and from the character of the bones themselves, I judged to have come from Triassic deposits. The sole attendant at that time present in the institution seemed as ignorant as myself regarding their history, as indeed he was of the other palæontological and zoölogical materials in the collection. His chief interest seemed to lie in a collection of the hearts of dead saints, or other worthies, preserved in alcohol and displayed in glass jars. These were fully labelled, and the attendant seemed at a loss to understand our total indifference to such objects, as he waxed eloquent in an attempt to set forth to us the principal features in the life history of each of the former owners of these now bleached and gruesome objects. What morbid interest or curiosity could prompt the friends of the departed to retain such repulsive souvenirs of their favorite dead is, I confess, quite beyond my understanding. Nevertheless the custom seemed to be not an unusual one in Catholic America, for I remembered that while visiting the church at the Recoleta both in Buenos Aires and Montevideo, we were shown similar portions of the anatomy of other favorite saints, preserved in spirits and placed on exhibition in the sacristy.

31 — Across the Patagonian plains in winter.

32 — Rio Paraguay.

FIGURES 31 AND 32—SEE OTHER SIDE

Asuncion, the capital of Paraguay, is of interest not alone from being the capital of that country, which to-day remains the most distinctive of all the South American Republics. It was the home of Francisco Solano Lopez, one of the greatest military spirits that South America ever produced. Not only was he an absolute dictator within his own country, but at one time he actually aspired to the proud distinction of being known as the Napoleon of South America, with authority over all the surrounding states. For a time, notwithstanding his limited resources, it did not seem at all improbable that he would succeed. He and his followers were possessed of such indomitable courage, that they were suppressed only after a protracted and bloody struggle, in which the combined armies of Brazil, Uruguay and Argentina were pitted against those of Paraguay. The latter were finally vanquished, but not until after the country was well-nigh depopulated.

Asuncion is a city of some forty thousand inhabitants, with fairly-well paved streets, poorly lighted, and with an indifferent tramway service, and quite destitute of carriages or other vehicles for getting about. The business portion of the city is ugly and monotonously uninteresting, but in the suburbs, where the better classes and the aristocracy dwell, there are beautiful villas with exquisitely laid-out and well-kept grounds.

As a people, the Paraguayans are unique among South Americans in that they have retained more of the original blood, manners, customs and language of the Indian people that formerly inhabited the region than those of any other country. The written and spoken language of the country is Paraguayan, the Spaniards, French and Portuguese never having been able successfully to introduce their own tongues. The upper classes are proud but not haughty. The women, though small, are remarkably handsome. The masses of the people are exceedingly poor, and, indeed, the country as a whole, at the time of my visit, seemed poverty-stricken. The soil is fertile, the wants of the people are few, and living correspondingly inexpensive. There is a single poorly-built and badly-equipped railway leading from Asuncion into the interior. Over this I travelled, and as we passed along, stopping at the not very frequent villages, I was struck with the great number of mendicants, who at the various stations fairly swarmed at the windows of the coaches, soliciting alms from the occupants. Most of these were suffering from the effects of syphilis or other loathsome diseases, usually already in an advanced stage. As an example of the poverty and the depreciated value of commodities, I may mention that when solicited by a fruit vender to purchase some oranges, at one of the small villages at which we stopped, on handing her a Paraguayan five-cent piece, equivalent at that time to just five eighths of one cent in our money, she gave me in return a bunch of thirteen large oranges, which were as fine in appearance and flavor as any I ever ate from the Indian River groves in Florida. They were woven together in one cluster by the stems in such a tasteful manner as to indicate that, if time were of any value whatever, the small pittance I had exchanged for them would scarcely repay the owner for her trouble and would leave her nothing for the fruit. Nevertheless, she seemed perfectly contented and quite pleased with the sale, and since I had allowed her to fix her own price, I had no compunction in the matter.

At San Bernardino I left the railway and went inland some miles to the quaint little town of Itaguay, where I remained for some time, and, although a tourist, I obtained an excellent room and good board at five pesos, sixty-two cents per day. If any of my readers should ever feel the need of a vacation spent amid quiet but comfortable and most interesting surroundings, at an exceedingly moderate expense, let me commend them to one of the country villages in Paraguay. During my short stay in that country I intentionally forgot everything and fell suddenly into the indolent manners and customs of the natives. Their ways became my ways, and if I walked or rode in the forests it was to satisfy not my mental, but physical desires and comforts. For once in my life I conquered the propensity for collecting which for years had been the one dominant and uncontrollable element within me. When I left Paraguay on my return voyage to Buenos Aires, I took with me just one natural history specimen — the shell of a large ground-snail belonging, I believe to the genus *Purpurea*, and picked up by a companion, while we were walking one day in the suburbs of Asuncion. These shells were very abundant and quite handsome. Some will doubtless say that I wasted a valuable opportunity in thus neglecting to collect in this most interesting region. But I do not think so, for I had come here for another purpose, and it had never been my nature to work in a half-interested manner. It was either to be absolute rest or strenuous and unrelenting work. I chose the former. Hence it is that I can say little of the fauna or flora of the country.

After my return to Buenos Aires I spent a few days in that city, sailing from La Plata on the Royal Mail Steamer "Clyde" for Rio de Janeiro, where I remained for five days, when I left for New York by the "Buffon," of the Lamport & Holt Line, and reached home on the sixteenth of August, 1899. Mr. Peterson was not long delayed, since he reached New York on September first, coming direct from Sandy Point with the "Capac," of the Grace Line. With him came the last of our collections, and our work in Patagonia, for a time at least, was finished.

We had undergone many hardships and made considerable sacrifices in order to accomplish the work. In many respects our success had far surpassed our most sanguine expectations, while we had signally failed in one most important feature of our work, which, however, I still hope to accomplish.

Almost every traveller in Patagonia has remarked and commented upon the deep impression made upon the mind by the vast expanse, aridity and solitude of the Patagonian plains. Darwin has attributed this to a certain realization of the fact that over vast regions they were at the time of his visit unknown and practically unknowable, as he then thought, on account of the supposedly inhospitable nature of these plains, thus leaving much of their true nature to the imagination. Hudson has taken exception to Darwin's explanation, holding that, while since Darwin's time the Patagonian plains have become quite as well known to travellers as many another region of the earth's surface of similar extent, yet that peculiar interest and impressiveness still attaches to them, and, once visited by the traveller, they ever remain as the most perfect, vivid and deeply engraved picture, more frequently and vividly recalled from among the retinue of paintings in his mental gallery than any of his other experiences, not excepting even those of our more impressionable childhood days. Hudson would seem to infer that the unusually distinct impression produced by the plains of Patagonia is due to and inherent in their monotony and lack of anything calculated to awaken in the mind of the traveller any real and sustained interest in matters other than the desolate solitude of his surroundings.

I am inclined to the opinion that both Darwin and Hudson correctly analyzed and set forth the reasons for their own impressions of Patagonia. Commencing with Chapter XIII. of the latter's "Idle Days in Patagonia," he says, "Near the end of Darwin's famous narrative of the

'*Beagle*,' there is a passage which, for me, has a very special interest and significance. It is as follows, and the italicization is mine: 'In calling up images of the past, I find the plains of Patagonia frequently cross before my eyes; yet these plains are pronounced by all to be most wretched and useless. They are characterized by only negative possessions; without habitations, without water, without trees, without mountains; they support only a few dwarfed plants. Why, then — *and the case is not peculiar to myself — have these arid wastes taken so firm possession of my mind?* Why have not the still more level, the greener, and more fertile pampas, which are serviceable to mankind, produced an equal impression? I can scarcely analyze these feelings, but it must be partly owing to the free scope given to the imagination. The plains of Patagonia are boundless, for they are scarcely practicable and hence unknown; they bear the stamp of having thus lasted for ages, and there appears no limit to their duration through future time. If, as the ancients supposed, the flat earth was surrounded by an impassable breadth of water, or by deserts heated to an intolerable excess, who would not look at these last boundaries to man's knowledge with deep, but ill-defined sensations?'

"That he did not in this passage hit on the right explanation of the sensations he experienced in Patagonia and of the strength of the impressions it made on his mind, I am quite convinced; for the thing is just as true to-day as of the time, in 1836, when he wrote that the case was not peculiar to himself. Yet since that date — which now, thanks to Darwin, seems so remote to the naturalist — those desolate regions have ceased to be impracticable, and, although still uninhabited and uninhabitable, except to a few nomads, they are no longer unknown." And on page 222, after a very graphic description of how, under certain circumstances, the mind of the individual living under the constant restraint of highly civilized conditions suddenly, through a momentary taste of adventure, lapses into a more natural and less artificial condition; he adds: "It was elation of this kind, the feeling experienced on going back to a mental condition we have outgrown, which I had in Patagonian solitudes; for I had undoubtedly *gone back;* and that state of intense watchfulness, or alertness rather, with suspension of the higher intellectual faculties, represented the mental state of the pure savage. He thinks little, reasons little, having a surer guide in his instinct; he is in perfect harmony with nature, and is nearly on a level, mentally, with the wild animals he preys on, and which in turn sometimes prey on him. If the plains of Patagonia affect a person in this way, even in a much less degree than in my case, it is not strange that they impress themselves so vividly on the mind, and remain fresh in memory, and return frequently, while other scenery, however grand or beautiful, fades gradually away, and is at last forgotten." Again, in describing a winter spent on the Rio Negro, he continues: "The valley alone was habitable, where there was water for man and beast, and a thin soil producing grass and grain; it is perfectly level, and ends abruptly at the foot of the bank or terrace-like formation of the higher barren plateau. It was my custom to go out every morning on horseback with my gun, and, followed by one dog, to ride away from the valley; and no sooner would I climb the terrace and plunge into the grey universal thicket, than I would find myself as completely alone and cut off from all sight and sound of human occupancy, as if five hundred instead of only five miles separated me from the hidden green valley and river. So wild and solitary and remote seemed that grey waste, stretching away into infinitude, a waste untrodden by man, and where the wild animals are so few that they have made no discoverable path in the wilderness of thorns. . . .

"Not once, nor twice, nor thrice, but day after day, I returned to this solitude, going to it in the morning as if to attend a festival, and leaving it only when hunger, thirst and the westering sun compelled me. *And yet I had no object in going.*" (Italics the present writer's.) In this last clause we have the key to the mental picture engraved by Patagonia on Hudson's mind. For, as the title of his very readable book, "Idle Days in Patagonia," implies, unlike Darwin, he was an idler, as must become apparent to any one who will take the trouble to read his book. Gifted with no mean literary talent and with no doubt a first-rate knowledge of ornithology, but somewhat deficient in a knowledge of general natural history, he found himself, through an unfortunate accident, compelled to remain for several months in a district not especially rich in ornithological material. Here he spent days, weeks, and even months, dreaming away his time in a most interesting region, adding little to our actual knowledge of its recent fauna or flora, and nothing concerning that past life, over the fossilized remains of which he and his horse must have stumbled daily, so abundant are they in the rocks of that region, as he climbed the slope. He spent each day in undisturbed dreams, while lying at ease on the clean sand beneath the shade and shelter of some friendly tree or bush that grew upon the summit.   If one but compares the aimless, idle manner in which, according to Hudson's own account, he passed the greater portion of his time, with the strenuous interest and exertion displayed by Darwin while travelling through the same region, the characteristics of the two men must appear quite distinct.   If Hudson found Patagonia monotonous, and so uninteresting as to compel him to lapse into a state of mental inactivity comparable, as he himself states, only to the normal mental condition of the savage, it was due to his personal equation rather than the uninteresting nature of the country.   Darwin found the region full of interest, with a wealth of material to stimulate the mind and awaken the energies, as is abundantly evidenced in his popular account of the voyage of the "Beagle," every page of which is delightful, and, although written sixty years ago, is to-day and will remain the greatest compendium of useful information extant concerning that region.

It is true that these plains are inhospitable, that over vast regions the curse of sterility is the one omnipresent characteristic, that the fauna and flora are meagre and little diversified, that the prevailing dull brown color imparted to the landscape by the scanty covering of dry and withered grass is monotonous and ill calculated to create enthusiasm in one of a highly artistic temperament, that, for the most part, these plains remain still uninhabited and are largely uninhabitable.   But do not these very facts lend to this region a certain interest?   If the plains of Patagonia are devoid of a single mountain, the level surface is relieved by a series of magnificent terraces and deep transverse valleys, traversed in some cases by noble rivers ; while the traveller who ventures far into the interior will be rewarded by the discovery of the remains of extinct craters and gigantic dikes filling huge fissures in the earth's crust, from which at a time in a not so remote past vast sheets of molten lava were poured forth over the surface of the surrounding country and now appear as lofty basaltic platforms, capping the higher tablelands of the interior over hundreds of square miles and intersected by a labyrinth of deep and almost inaccessible cañons, which, by reason of their rugged and picturesque nature, supply most of the essential features of an excessively mountainous region, save that in this case the traveller over the plains is forced to descend rather than ascend in order to observe the full effects of nature's handiwork.   For in this instance she has departed from her usual custom and has placed her gallery in the cellar, as it were, though as usual the walls are hung with many a striking piece, executed with a boldness of design and in such harmonious colors as are the envy and aspiration of every painter of landscapes.

That would be a dull mind, indeed, which could contemplate without interest these vast, almost limitless plains, unsurpassed elsewhere in the world, covered with a bed of shingle that is nowhere else equalled in extent, and the origin of which has yet to be satisfactorily explained, underlaid by several thousand feet of sedimentary rocks, certain strata of which are literally filled with the remains of an animal life which has so completely and entirely disappeared from our earth, that to-day not only genera and species, but whole families and entire orders are no longer represented in the living fauna of this or any other part of its surface. I cannot conceive that it were possible for Mr. Hudson to have avoided an interest in some of these phenomena, or an attempt at an explanation of them. Concerning my own impressions of Patagonia, unlike Mr. Hudson, I cannot complain of feeling the want of any stimulus to either mental or physical activity. During the three years passed in that region I was incessantly, and, I think, profitably employed, and at present my chief regrets are that I had to leave so many interesting problems unsolved and that I am so deficient in the literary talents possessed by Mr. Hudson, which would have enabled me to place before my readers the results of my work and observations in the delightful style employed by that author in "The Naturalist in La Plata," "Idle Days in Patagonia," and a number of other equally interesting and charming books. Perhaps on account of a life-long familiarity with our own western plains, quite as extensive and frequently as barren as those of Patagonia, the latter did not impress me as they evidently have impressed other travellers. Though I suffered much from the inhospitable nature of the climate of Patagonia, I am forced to confess to a certain very warm attachment to that country, and I know of no other one thing that would cause me more pain than to be forced to abandon all hope of ever again visiting the region for the purpose of continuing and, if possible, completing my investigations.

CARNEGIE MUSEUM, February 1, 1902.   J. B. HATCHER.